T0292011

It's in Your DNA

It's in Your DNA

From Discovery to Structure, Function, and Role in Evolution, Cancer, and Aging

Dr. Eugene Rosenberg
Department of Molecular Microbiology and Biotechnology,
Tel Aviv University, Tel Aviv, Israel

With drawings by
Avshalom Falk

Academic Press is an imprint of Elsevier
125 London Wall, London EC2Y 5AS, United Kingdom
525 B Street, Suite 1800, San Diego, CA 92101-4495, United States
50 Hampshire Street, 5th Floor, Cambridge, MA 02139, United States
The Boulevard, Langford Lane, Kidlington, Oxford OX5 1GB, United Kingdom

Library of Congress Cataloging-in-Publication Data
A catalog record for this book is available from the Library of Congress

British Library Cataloguing-in-Publication Data
A catalogue record for this book is available from the British Library

ISBN: 978-0-12-812502-1

For information on all Academic Press publications visit our website at https://www.elsevier.com/books-and-journals

Publisher: Sara Tenney
Acquisition Editor: Kristi Gomez
Editorial Project Manager: Pat Gonzalez
Production Project Manager: Karen East and Kirsty Halterman
Designer: Matthew Limbert

Typeset by Thomson Digital

For Ilana, my partner in science and life.

Contents

Preface

The mission of this book is to share with the interested reader the pleasure of understanding one of the greatest achievements of science—uncovering the mystery of DNA. To fully appreciate this achievement, I have strived to explain in a simple but scientifically correct manner the key experiments and concepts that led to our current knowledge of what DNA is, how it works and the increasing impact it has on our lives. Although the book was written for a layperson to understand, I also kept in mind students, teachers, and young scientists.

In addition to emphasizing the observations and reasoning behind each novel idea and the critical experiments that were performed to test them, I have provided a brief sketch of the lives and personalities of key scientists, many of whom I knew personally. The importance of DNA research to science and medicine is reflected by the fact that 27 Nobel Prizes have been awarded for research on DNA. The prizes were awarded to scientists from 12 countries (Appendix). Among the pioneers of DNA research that will be discussed in this book are four extraordinary women, who have become feminine icons: Rosalind Franklin, whose X-ray pictures of DNA made possible the discovery of the double-strand helix structure of DNA; Barbara McClintock, whose discovery of "jumping genes" gave rise to genetic engineering; Lynn Margulis, who put forth original ideas on evolution that were initially controversial but are now accepted; and Elizabeth Blackburn, who discovered that shortening the ends of the DNA molecule contributes to aging. Telling a science story from the point of view of the scientist enables the reader to imagine themselves in that position.

In narrating the DNA story, I have used the historical or case-study approach, in which I describe the seminal ideas in DNA research from their conception to their latest development, emphasizing the observations and reasoning behind each concept and the critical experiments that were performed to test them. In this way, I intend to stimulate the reader's innate sense of scientific inquiry. Learning how someone made an important discovery can be as interesting as the discovery itself. In the DNA story, the journey matters. For it is precisely in the substitution of evidence for dogma, as a basis for belief, that science has made its greatest offering.

The historical approach also leads the reader to appreciate how science progresses. An important discovery leads to a new question. Often the question is put in the form of a hypothesis that can be tested experimentally. The results of the experiment can lead to a new concept, which again provokes another question. Revolutions occur when disparities or anomalies arise between theoretical expectation and research findings that can be resolved only by changing fundamental rules of practice. These changes can occur suddenly: in a relative instant, the perceived relationships among the parts of a picture shift, and the whole takes on a new meaning. Examples that are included in this book include Darwin's evolutionary theory, Mendel's gene concept, Avery's demonstration that DNA is the genetic material, the helical model of DNA, the one gene–one enzyme concept, and more recently, the key role of symbiotic microorganisms in adaptation, behavior, development, metabolism, and evolution of plants and animals, including humans.

Throughout the book, I have attempted to strike a balance between theoretical and applied aspects of DNA research. Scientific progress is motivated by both a desire to know and a desire to use this knowledge. Nowhere is this more evident than in DNA biology. The ability to manipulate DNA has led to genetically engineering microbes to produce valuable drugs, such as the human hormone insulin, human growth hormones to prevent dwarfism, blood-clotting factors for treating hemophilia (the "bleeding disease"), and vaccines. Genetic engineering is used in agriculture to produce plants with beneficial characteristics, such as herbicide resistance, improved shelf life, disease resistance, stress resistance, and resistance to insects. The DNA engineered crop that probably has the greatest potential to improve human health worldwide is Golden rice, created to fight vitamin A deficiency, which affects 250 million people around the world and can cause blindness and even death.

Genetic engineering has also been performed on animals. Although most of the genetically modified animals were produced for research purposes, some were engineered to enhance production or food quality traits. Two of the most interesting are genetically modified salmon for use in the aquaculture industry to increase the speed of development and potentially reduce fishing pressure on wild stocks and dairy cows genetically engineered with human genes to produce milk that would be the same as human breast milk. This could potentially benefit mothers who cannot produce breast milk but want their children to have breast milk rather than formula.

Genetic engineering of humans is in its infancy. Recently, the British House of Commons approved a technique of creating babies from three people, the mother, the father, and a female mitochondrial donor. The technique was developed to help women with mitochondrial disease have healthy babies. Eggs are collected from a potential mother with damaged mitochondrial DNA and a female donor with healthy mitochondria. The nucleus which contains the vast majority of the genetic material is then removed from both eggs. The mother's nucleus is inserted into the donor egg, which can then be fertilized by sperm from the father. It results in babies with 0.1% of their DNA from the second woman and is a permanent change that would be passed down through the generations.

I hope this book stimulates interest in DNA and provides the necessary background for a greater appreciation of the increasing number of discoveries that are making an impact on our way of life.

Acknowledgments

I have several people to thank. My colleagues, Ed Kosower, David Gutnick, Gerry Cohen, Irun Cohen, Moshe Meverich, Isaak Witz, and John Ringo read parts of the book and provided valuable suggestions and up-to-date references in their specialties. I developed the hologenome concept of evolution in collaboration with my wife and scientific partner, Ilana Zilber-Rosenberg. My daughter, Stephanie Rotem, read large parts of the book and offered many suggestions for clarification from the advantage point of an interested layperson.

Avshalom Falk drew the amusing and informative cartoons that are spread throughout the book. Avshalom is a professional microbiologist who has the ability to see humor in science and the talent to express it in the form of artwork.

I was blessed by having excellent teachers during my formative years, Dan Atkinson, Syd Rittenberg, and Ralph Martinez at UCLA, Steve Zamenhof and David Rittenberg at Columbia University, and C.B. van Niel at Stanford University. When I was a graduate student, Zamenhof told me that if I really understood a subject I should be able to write it in a manner that my mother-in-law could understand it. I tried to follow his advice in this book.

I also thank the editors of Elsevier, especially Kristi Gomez, Patricia Gonzalez, Karen East and Kirsty Halterman, who recognized the value of the book and helped bring it to fruition.

Acknowledgements

[illegible]

Prologue

In 1974, 18-year-old James Bain was convicted of breaking and entering, kidnapping, and rape of a 9-year-old boy in Polk County, Florida, and sentenced to life in prison. The story is a long and tragic one with many mistakes and misjudgments. The 9-year-old boy was kidnapped from his home on the night of March 4th, dragged to a nearby field, and raped. When the victim returned home, he was wearing only a white t-shirt and jockey underwear. The police collected his underwear and sent it to the FBI for serological testing. Bain became a suspect after the victim described his assailant as a 17- or 18-year-old African–American man named "Jimmy" having bushy sideburns and a mustache. The boy also told the police that he remembered seeing a red motorcycle; a similar one of this description was owned by Bain. In addition, the victim's uncle stated that the man he was describing sounded like James Bain, who he knew as a student at the high school where he was assistant principal. After his uncle mentioned the name, the victim adopted Bain as the rapist. When the police arrived at Bain's house just after midnight on the night of the crime, Bain was at home with his sister where he had been all evening. Although James Bain had an alibi, did not confess to the crime, and maintained his innocence continuously; the police arrested him.

During the trial, the defense argued that Bain had no criminal record and could not have been the rapist because he was home at the time, 2 miles from the scene of the rape. Four family members testified that Bain was home watching television with his sister during the specified time. However, family members are known to be poor alibi witnesses because juries believe they could be biased, saying only what is good for the family member. Later Bain said: "But (the jury) didn't pay that no attention. I think they mainly convicted me when the victim stood up in the courtroom and pointed at me as the perpetrator. He was crying and everything in the process."

The State's case was also based on blood tests on the jockey underwear of the victim performed pretrial by the FBI. At the time, serology was considered the most technologically advanced procedure for identifying criminals. The tests verified that the rapist did deposit semen on the victim's underwear and sperm heads were seen on the underwear. FBI analyst, William A. Gavin, testified that the semen on the underwear was blood group B, that Bain's blood group was AB (with a weak A), and that Bain could not be excluded as the depositor of the semen. The underwear itself was admitted at trial as State's Exhibit #1. Contrary to the

FBI testimony, a defense expert, Dr. Richard Jones, testified at the trial that Bain's blood group was AB (with a strong A) leading to his conclusion, contrary to the FBI's, that Bain could not have deposited the semen on the victim's underwear. Despite the conflicting testimony on the serology report, the jury convicted Bain of all three charges and sentenced him to life in prison.

During the 35 years that Bain was incarcerated in six different prisons, he became a master welder, learned to play chess, and studied various subjects in school for 27 of his 35 years behind bars. Bain grew close with many other prisoners but often waited years to tell them he was innocent. "I had to know someone before I told them. Anything could go wrong. If they don't believe you, they could get mad. Y'all have a fight, and someone winds up dead," he said. In spite of this mistrust, from his first day behind bars, he befriended older inmates with some level of legal knowledge to help him write requests to the court for transcripts of his trial—a right granted by the Supreme Court. But he neither received them nor heard back from the court. Eventually, Bain learned about DNA evidence.

DNA evidence was used for the first time in 1988 to convict a murderer and in 1989 to pardon a convicted person who was serving time in a US prison. Since then, DNA evidence has been used to exonerate more than 300 persons either on death row or serving lengthy prison sentences. When Bain was informed of the novel DNA technology, he submitted handwritten motions 5 times seeking DNA testing, but he was denied each time. On the fifth time that James was denied his request for testing, the appeals court wrote that he had waited too long. It wasn't until Florida passed a new law that allowed cases to be reopened for DNA testing that his fifth and final rejection was overturned and early in 2009 the Innocence Project of Florida stepped in to assist Bain. The Innocence Project is a nonprofit legal organization that is committed to exonerating wrongly convicted people through the use of DNA testing. The State Attorney agreed to the DNA testing and the victim's underwear worn during the rape in 1974 was sent to DNA Diagnostics Center, a private laboratory in Fairfield, Ohio. Fortunately, the underwear had been stored in plastic bags in A/C rooms, so the DNA was accidentally preserved. Testing the DNA of sperm found on the underwear excluded James Bain as the donor of the sperm, confirming that someone other than Bain raped the victim. According to the Innocence Project Legal Organization, misidentification from an eyewitness accounts for 75% of all convictions later overturned by DNA evidence.

It has been known for many years that DNA testing is valuable to the justice system because it is extremely sensitive and unbiased. Only a few molecules are required to identify the individual that deposited the DNA. Furthermore, unlike witness identification of a suspect, the test is not prejudice to race, gender, or economic status.

On December 17, 2009, James Bain was declared innocent and released from prison. He walked out of the courthouse with members of his family and his legal team. Bain had spent 35 years in prison—the longest time served by innocent man eventually freed using DNA evidence.

James Bain. *(From https://www.innocenceproject.org/cases/james-bain/. Used with permission from Florida Innocence Project.)*

Under Florida's wrongful incarceration statute, Bain received compensation from the state of Florida for his time in prison—about $1.7 million, $50,000 for every year he spent in prison. A few months after he was released, he said: "Part of me went into hibernation in prison, but every day, I'm waking up a little more." When taken to a 3-D screening of Avatar, he commented, "Most definitely amazing," but what amazed him more than the movie, he says, was all the iPhones and BlackBerrys flashing in the audience during the previews, "Like being on an airport runway, with rows of small lights everywhere."

He thinks about traveling a lot, he says. But for now, he's home with family, reveling in the company of his mother, siblings, nieces, and nephews. He sleeps with a fan on to feel the movement of air. He sits with his hands apart to remind himself there are no handcuffs. He eats with a fork, knife, and spoon. "I'm rusty, but I'm coming back," he says.

Speaking to students on his wrongful conviction experience at the University of South Florida on February 24, 2016, Bain told listeners: "I tried and tried, ladies and gentlemen, to get my case heard. But each time, it was a failure. Each time."

The exoneration of James Bain is just one example of how knowledge about DNA has affected the lives of people in the recent past and will do so to an even greater extent in the future. DNA fingerprinting is used not only in criminal investigations but also has revolutionized maternity/paternity testing, forensics, and identification of disaster victims.

Basic and applied research on DNA has led to enormous progress in biology and medicine. Deoxyribonucleic acid, or DNA for short, has been playing a central role in all organisms, from simple microbes to man, for more than 3 billion years. However, it is only in the last century that we have begun to understand what DNA is and how it operates to determine what we are, and how this information is transmitted to our children. Biology only makes sense in terms of evolution, and changes in DNA are at the core of evolution. DNA provides the instructions for producing all life forms. We now have the tools to read these instructions and begin to understand the sequence of changes in DNA that led to the evolution of *Homo sapiens* as well as other animals and plants.

An interesting example of how DNA technology can teach us about the ancient history and evolution of modern humans is the analysis of DNA from a 45,000-year-old bone of a Neanderthal man found in Siberia. The analysis indicates that 1%–4% of the DNA of anyone living outside Africa today is Neanderthal in origin. These findings reveal that Neanderthals interbred with ancestors of modern humans when modern humans began spreading out of Africa. Another surprising discovery based on a detailed analysis of the DNA of the Neanderthal man is that he had red hair and pale skin.

Recent results have shown that many important evolutionary innovations involved acquiring genes from microbes, rather than building them from scratch. In fact, all animals and plants live in symbiotic association with abundant and diverse microbes. Humans, for example, contain as many microbial cells as human cells and more than 400 times more microbial genes than human genes. The title of this book, "*It's in Your DNA*" refers to both your human and microbial DNA. Your symbiotic microbes play an important role in your development, behavior, general health, and ability to adapt to changes in your diet.

DNA technology has revolutionized modern medicine. Many therapeutic drugs, such as insulin, human growth hormone, and human blood clotting factors, are now produced using DNA technology, making them much less expensive and safer. For example, prior to the development of DNA technology, human blood clotting factors were produced from donated blood that was inadequately screened for HIV. Insulin was extracted from the pancreas glands of cattle, pigs, and other farm animals. Animal-derived insulin is different from human insulin, and therefore produced allergic reactions in some patients. Human growth hormone was manufactured by extraction from the pituitary glands of cadavers, as animal growth hormones have no therapeutic value in humans. Production of a single year's supply of human growth hormone for one patient required up to 50 pituitary glands, creating significant shortages of the hormone.

DNA-based methods are used for rapid diagnosis and monitoring of hundreds of maladies, from determining the specific microbial strain that causes an infectious disease to detection of cancer. Seventy percent of clinical decision-making is now based on diagnostic test results. The pharmaceutical industry estimates that annual revenue from the sale of diagnostic kits exceeds US $200 billion. More importantly, rapid and accurate diagnostic tests save millions of lives.

Two of the major medical problems today are cancer and the aging process. As I will explain in this book, deleterious changes in DNA are at the core of both problems. The challenge is to discover ways to prevent accumulating DNA damage and possible methods to reverse the process in order to provide for longer and healthier lives.

The impact of DNA in our life can be illustrated from a few of the many recent *New York Times* headlines:

MPs Say Yes to Three-Person Babies
Rise in Overturned Sexual Abuse Convictions From DNA Testing
A Genetic Entrepreneur Sets His Sights on Aging and Death
Genetic Manipulations to Avoid Devastating Diseases in Babies
Bringing Extinct Animals Back to Life is Really Happening
Genetically Modified Foods: Questions Remain
I Like Your Genes: People More Likely to Choose a Spouse With Similar DNA
Genes That Might Determine Whether You Have Straight, Curly or Wavy Hair
How Far Would You Go to Modify Yourself Using the Latest DNA Technology?
How DNA Forensics Could Identify Lost Nigerian Girls

All of these subjects are part of the exciting story of DNA that I wish to share with you. The story involves creative scientists, who are often interesting people, brilliant experiments, which I will endeavor to describe in a clear and accurate manner, and an outcome that gives us some understanding of the most basic questions in biology: Where do we come from? What are the causes of cancer and aging? What does the future hold for humans? Let me assure you at the outset that the DNA story is so interesting that even without adding spice, it will taste good.

Chapter 1

The Beginning

The Past—the dark unfathomed retrospect!
The teeming gulf—the sleepers and the shadows! The past!
The infinite greatness of the past!
For what is the present after all but a growth out of the past?

—Walt Whitman, Passage to India in *Leaves of Grass*

The past is never dead. In fact, it's not even past.

—William Faulkner in *Requiem for a Nun*

Theoretical physicists hypothesize that the universe began with a colossal explosion often referred to as the *Big Bang* theory, approximately 13.8 billion years ago. The key idea is that the universe is expanding. Consequently, the universe must have been denser and hotter in the past. In particular, the *big bang* theory suggests that at some moment all matter in the universe was contained in a single point, which is considered the beginning of the universe. After the initial expansion, the universe cooled sufficiently to allow the formation of subatomic particles, including protons, neutrons, and electrons. The gases and dust from that explosion produced the earliest generation of stars, and over a period of billions of years, the stars exploded, and their debris formed other stars and planets. Our solar system was formed in this way 4–5 billion years ago. During the next billion years, the molten Earth cooled, forming a hardened outer crust.

About 3.8 billion years ago, earth's atmosphere consisted of gases, such as hydrogen (H_2), nitrogen (N_2), hydrogen sulfide (H_2S), methane (CH_4), and water (H_2O). As the temperature decreased, water vapor condensed causing millions of years of torrential rains, during which time the oceans formed. Gases and water from the earth's core came to the surface through volcanoes. Ultraviolet radiation bathed the earth, and the simple compounds interacted with one another to form more complex organic molecules, including the building blocks for life. Organic molecules are those containing carbon and that are typically found in living systems.

How organic molecules formed the first living cells, that is, the origin of life, remains one of the most challenging unsolved problems in biology. The best fossil evidence indicates that microorganisms first appeared on Earth about 3.8 billion years ago and the first multicellular creatures about 1 billion years ago. Thus,

It's in Your DNA. http://dx.doi.org/10.1016/B978-0-12-812502-1.00001-9

microbes were the only living organisms on this planet for more than 2 billion years. During this time, the microbes evolved most of the processes we associate with life, stored this information in the form of DNA, and passed it on to plants and animals. The vast majority of DNA information in humans (*Homo sapiens*), which first evolved about 200,000 years ago, is remarkably similar to the DNA of microbes.

It is truly amazing that from the *big bang* to atoms and molecules, to stars and planets, to microbes and the evolution of more complex animals and plants, and finally to humans, we can now begin to understand these processes. This is the DNA story.

Like a river, every story, even a scientific story, has a source, a beginning. If you ask the question: who discovered DNA? Most educated individuals, including many scientists, will answer "Watson and Crick, of course!" However, the story of DNA actually begins with the Swiss scientist Johann Friedrich (Fritz) Miescher in 1871.

Miescher was born in Basel, Switzerland, in 1844, the eldest of five sons of Friedrich Miescher-His, professor of pathologic anatomy at Basel University, and a successful practicing gynecologist. The Miescher family was well-respected and part of the intellectual elite in Basel. Friedrich's uncle, Wilhelm His, who lived in the same house as the Mieschers, was professor of anatomy and physiology, distinguished for his work in embryology and histology. He had a life-long influence on his nephew. The young Friedrich was an excellent student despite his shyness and a hearing handicap. Initially he wanted to be a priest, but his father opposed the idea and Miescher entered medical school. At the time of his graduation in 1868, he wrote a long letter to his father, discussing his career plans. He wanted to be a practicing physician, but because of his hearing difficulties, he chose ophthalmology, where listening through a stethoscope would not be needed. On the other hand, he had a great desire to work in basic research. He wrote: "It was only in the lectures on physiology that the entire splendor of research on organic matters became apparent to me." He therefore proposed a compromise: he would practice ophthalmology and in his spare time do research. Miescher's father showed the letter to Fredrich's uncle, Wilhelm His, who instantly saw that the compromise solution would not work. He proposed that "in view of the considerable mental talents which Fritz has," he should enter the career he found most appealing, that of research in physiology. Someone as eminently theoretical in nature as Miescher would find satisfaction only in scientific research. Miescher, who idolized his uncle, followed his advice and his father agreed.

Miescher chose to be trained as a scientist in the laboratory of the eminent biochemist Felix Hoppe-Seyler at Eberhard-Karls-University in Tübingen, Germany. Hoppe-Seyler was one of the pioneers of what was then referred to as physiological chemistry, a new field aiming to unravel the chemistry of life. Hoppe-Seyler performed seminal work on the properties of proteins, most notably hemoglobin (which he named), and introduced the term proteid which later became protein. Hoppe-Seyler's expertise and research interests were closely aligned with Miescher's aims and his laboratory, housed in Tübingen's Castle, proved to be a stimulating place for Miescher to work.

At 372 m, Tübingen's Castle offers a magnificent view of the Neckar and Ammer Valleys. The castle dates back to 1078 and is of renaissance construction with four wings and a round tower. The rulers of Tübingen, who were promoted to Counts Palatine in the 12th century, lived in the castle until 1342 when they sold it to the Counts of Württemberg. Beginning in the mid-18th century, the university acquired its first rooms in the castle and in 1816 the King of Württemberg, Wilhelm I, transferred ownership of the castle to the university. The university library of nearly 60,000 bands was housed in the hall of knights, and an astronomical observatory was housed in the northeast tower. Hoppe-Seyler's laboratory occupied the royal laundry room in the basement; he found Miescher space next door, in the old kitchen.

At their first meeting, Hoppe-Seyler proposed to Miescher, then aged 23, that he perform research on the composition of lymphoid cells—white blood cells. Hoppe-Seyler was aware from microscopic examinations that white blood cells have a large nucleus, so that examining these cells might reveal information on the chemistry of the nucleus. However, Miescher's initial attempts to isolate white blood cells from lymph glands proved difficult. He was unable to obtain enough cells for analyses. Miescher then turned to surgical bandages discarded in a nearby surgical clinic where soldiers were stationed. The bandages were an excellent source because lymphoid cells are abundant in pus from infections. The problem he now faced was washing the cells off the bandages without damaging them. By trial-and-error, Miescher found that a particular salt solution containing sodium sulfate was effective. After filtering the solution to remove residual cotton fibers, he allowed the cells to settle at the bottom of a beaker. Examination in the microscope indicated that the cells were intact and free of contaminating materials. By this simple method, Miescher obtained large enough quantities of the cells for analysis (Fig. 1.1).

FIGURE 1.1 Friedrich Miescher (1844–95).

One of the first experiments Miescher carried out with the isolated white blood cells was to treat them with gastric juice (we now know that gastric juice contains an enzyme, pepsin, which digests protein). Today, one would simply buy pepsin from one of the many companies that produce the enzyme. However, Miescher had to obtain the gastric juice himself by extracting the juice from pigs' stomachs. As observed in the microscope, the treatment of the white blood cells with the gastric juice dissolved the cytoplasm of the cell, leaving only a fine gray precipitate consisting of shrunken nuclei. Since the material was isolated from nuclei, he called it "nuclein."

Nuclein had one property that was typical of proteins; it dissolved in mild alkali and could be precipitated with cold acid. However, when Miescher subjected nuclein to elementary analysis, one of the few chemical tests available at the time, it revealed a weight composition that was different from any cell component known at that time: 14% nitrogen (N), 2% sulfur (S), and 6% phosphorus (P). The high phosphorus content was particularly noteworthy since proteins lack that element.

When a scientist, or for that matter anyone, is confronted with an unexpected finding, one has the choice of ignoring it and moving on, or trying to explain it. Miescher was shy, but he did not lack self-confidence. He did what any good experimental scientist would do—he repeated the experiment, following the exact procedure he had written in his notebook. When he achieved the same results, he must have had a eureka moment, realizing that he had made an important discovery—nuclein was a novel material present in the nucleus of white blood cells.

The final studies on nuclein in Tübingen were made in the summer of 1869, after which Miescher returned to Basel in his native Switzerland to write a manuscript on his research. The manuscript, dated "Basel, October 1869," was sent to his mentor Hoppe-Seyler, who was skeptical about the rather revolutionary findings of a beginner, especially the unprecedented high concentration of phosphorus. Accordingly, he decided to repeat the experiments himself, and he published Miescher's paper "Über die chemische Zusammensetzung der Eiter-zellen" (On the Chemical Composition of Pus Cells) in 1871 only after he had verified the results. Hoppe-Seyler wrote in a footnote to the paper: "I have to emphasize that in all points as far as I have examined Miescher's statements I have to confirm the latter fully." The ability to repeat an experiment is fundamental to science as it provides the strong foundation for further research. In addition to confirming Miescher's results on white blood cells, Hoppe-Seyler demonstrated the presence of nuclein also in the nucleus of red blood cells of birds and snakes. (In contrast to mammals, bird and reptile red blood cells have nuclei.)

Miescher was appointed to the newly created chair of professor of physiology at Basel University in 1872 at the exceptionally young age of 28, after receiving the highest recommendations from Hoppe-Seyler. Miescher started with very poor facilities, his laboratory consisting of a converted corridor and helped only by a quarter-time technician. His old friend, Alfred Jaquet, after Miescher's death, commented that, when Miescher took over the professorship in Basel, aware of his great new responsibilities, he tried to make up for his youth and inexperience by redoubling his research and teaching efforts. He continued to work on nuclein for the rest of his career, showing that all cells examined contained nuclein. Salmon sperm was a particularly good source of the material because they were simple independent cells, known from previous microscopic studies of their anatomy to consist almost entirely of nuclei. Moreover, on the practical side, the salmon migration up the Rhine and the consequent large-scale salmon fisheries made this material easily available in large quantities. However, none of Miescher's discoveries in Basel were as significant as his milestone discovery of nuclein when he was in his early 20s.

Much of what we know of Miescher's personal life comes from a monumental work by Wilhelm His, in which he describes Miescher's life, appending a collection of Miescher's correspondence with family and colleagues, in addition to including a selection of his scientific papers. Miescher married Marie Ann Rusch in 1878; they had three children. Miescher's personality was described by one of his students: "He appeared to be insecure, restless, and introverted no doubt because he was hard-of-hearing and myopic. The impression he gave was of a person completely taken up by his inner mental activity, without contact with the outer world. However, when called upon, he was always ready to help others. As a lecturer, he was difficult to understand, fidgety, in constant motion, with little contact with his audience" (Fig. 1.2).

In 1885, Miescher contracted a lung inflammation, possibly because of the long hours spent in his ice-cold laboratory. He neglected his ill health to the

FIGURE 1.2 **Miescher found salmon sperm to an excellent source of nuclein.**

extent of not taking any nourishment for long periods. In 1894, he was diagnosed as suffering from tuberculosis. He died of the disease in a sanatorium in Davos, Switzerland, in 1895, at age 51. One of his colleagues wrote to him shortly before his death, "I know what it is to give up well-loved, hopeful work. Sad as it is, there remains for you the satisfaction of having completed immortal studies in which the main point has been the knowledge of the nucleus; and so, as men work on the cell in the course of the following centuries, your name will be gratefully remembered as the pioneer of this field." He could not have been more correct.

Miescher's research on nuclein was continued by his successors. When all protein was finally removed from nuclein, it became clear it was an acid, and it was then referred to as nucleic acid.

Just prior to Miescher's publication on his findings on nuclein (1871), which eventually led to the discovery of DNA, Charles Darwin published his book "Origin of Species by Natural Selection" (1859), and the Austrian monk Gregor Mendel published the laws of inheritance (1866). It is interesting to compare the circumstances of these discoveries and their impacts.

Darwin's theory of evolution by natural selection, derived not from experiments but from observations on animals and plants made during his 5-year voyage to the coasts of South America and the Galapagos Islands, was immediately acknowledged as a major breakthrough. Thomas Huxley in the April 1860 issue of the Westminster Review hailed the book as, "a veritable Whitworth gun in the armory of liberalism" promoting scientific naturalism over theology, and praising the usefulness of Darwin's ideas. Huxley compared Darwin's achievement to that of Copernicus in explaining planetary motion. Darwin's books on evolution have remained the foundation of the discipline of evolutionary biology for more than 150 years.

Mendel's experiments on the transmission of hereditary traits in common pea plants, carried out in the garden of a monastery, were published in 1866 in a minor journal and were seen as essentially about hybridization, or blending, rather than inheritance of traits. It had little impact and was cited only 3 times over the next 35 years. In 1900, three researchers, each from a different country, rediscovered Mendel's laws of inheritance. Mendel's results were quickly replicated; biologists flocked to the theory, and Mendelism became the foundation of genetics.

Miescher's discovery of nucleic acid was published in a reputable journal and immediately accepted as a novel natural material, largely because the discovery came out of the laboratory of the famous Hoppe-Seyler. However, since no function was ascribed to nucleic acids, it was not considered an exceptional discovery at the time. Only in the middle of the 20th century, when it was shown that genes are composed of nucleic acids, was the importance of his discovery realized. That was the time when the disciplines of genetics, evolutionary biology and nucleic acid biochemistry merged and gave rise to molecular genetics.

In the next chapter it will be shown that there are two types of nucleic acid: deoxyribonucleic acid (DNA) and ribonucleic acid (RNA). Miescher's nuclein was DNA. We now realize that DNA contains the blueprint for life and is the basic material of genes. It is inherited from cell to offspring and from parents to children and changes in the DNA are the raw materials for evolution.

What were the personality traits that allowed Darwin, Mendel, and Miescher to make their revolutionary discoveries? Although there is no single mold for building great scientists, certain traits seem be common to the major innovators—independence, stubbornness, self-confidence, and the ability to concentrate for long periods and work hard on subjects that interest them. These characteristics appear to be present in most, if not all, the key scientists that contributed to solving the mystery of DNA, what it is and how it works.

The trait of concentration, not allowing yourself to be detracted from your goal, reminds me of a humorous event I observed when I was a graduate student. My mentor, Professor Stephen Zamenhof, was invited to a cocktail party at Yeshiva University in honor of the Israeli Nobel Prize winner in literature, Shai Agnon, and asked me to accompany him. Agnon immediately recognized Zamenhof and began speaking to him in the language they both enjoyed, Yiddish. The host of the party, Samuel Belkin, President of Yeshiva University, kept trying to introduce Agnon to members of his Faculty. Finally, when Belkin tapped Agnon on the shoulder, Agnon turned and asked Belkin, a distinguished Torah scholar, "Why did it take God only 6 days to make the Earth? When Belkin had no answer, Agnon said: "Nobody interrupted him," and returned to his focused conversation with Zamenhof (Stephen Zamenhof was the nephew of Ludwik Lazarus Zamenhof, the creator of Esperanto, the world's most successful constructed language).

NOTES AND REFERENCES

1 Roos, M., 2008. Expansion of the Universe—Standard Big Bang Model. In: Engvold, O., Stabell, R., Czerny, B., et al. (Eds.), Astronomy and Astrophysics. Encyclopedia of Life Support Systems. UNESCO.

1 Walt, W., 1998. Passage to India in Leaves of Grass (1871). In: LeMaster, J.R., Kummings, D.D. (Eds.), Walt Whitman: An Encyclopedia. Garland Publishing, New York, NY.

1 Thoreau, H.D., 2011. A Week on the Concord and Merrimack Rivers (1849). Princeton University Press, Princeton, NJ.

4 The all-important phosphorus analysis, which led to the discovery of nuclein consisted in combusting the phosphorus-containing substance with sodium nitrate and carbonate, dissolving the residue in nitric acid and precipitating with ammonium molybdate. The precipitate of ammonium phosphomolybdate was dissolved in ammonia, re-precipitated with magnesium sulfate and weighed as magnesium pyrophosphate. All the methods used at the time were gravimetric and therefore laborious and time-consuming.

4 Friedrich, K., 2000. Nucleic acids revelation delayed by a sceptic. Nature 403, 478.

4 Friedrich, M., 1871. Ueber die chemische Zusammensetzung der Eiterzellen. Medizinisch-chemische Untersuchungen 4, 441–460.
5 His, W., 1897. Die Histochemischen und Physiologischen Arbeiten von Friedrich Miescher. F.C.W. Vogel Publishing, Leipzig.
6 Mendel, G., 1901. Versuche über Pflanzen-Hybriden. Verh. Naturforsch. Ver. Brünn 4, 3–47 (1866). In: English, J.R. Hortic. Soc. 26, 1–32.
6 Huxley, T.H., 1860. Darwin on the origin of Species. Westminster Review. pp. 541–570.

Chapter 2

Chemistry of DNA

Research is formalized curiosity. It is poking and prying with a purpose.

—Zora Neale Hurston, American writer and anthropologist, 1942

I placed some of the DNA on the ends of my fingers and rubbed them together. The stuff was sticky. It began to dissolve on my skin. It's melting -- like cotton candy. Sure, that's the sugar in the DNA. Would it taste sweet? No. DNA is an acid.

—Timothy Ferris, in Richard Preston's biographical essay on Craig Venter, *The Genome Warrior* (originally published in *The New Yorker* in 2000)

To a large extent science is built vertically: the results from one set of experiments often provide the foundation for the next. In this way, science progresses, yielding a greater understanding of the natural world. Throughout this book I will show the progression of experiments that eventually led to our current understanding of the structure of DNA and how it functions in the transmission and expression of genetic information. After Miescher's discovery of nucleic acid, biochemical research on nucleic acid focused on determining its chemical structure. Rather than simply presenting our current knowledge on the structure of DNA, it will be by far more interesting and informative to learn about that through the observations and experiments which led to this knowledge. In this way one can obtain some insight into how a structure as complicated as a DNA molecule can be elucidated, while at the same time learning how DNA does what it is supposed to do—control cellular and whole body activity in addition to being the molecule in charge of inheritance.

Chemistry is a subject that most nonscientists avoid, at least in part because it is generally taught in a boring manner in schools. But there is a beauty in chemistry, the ability to determine the detailed structures of materials without actually seeing them. Telling the story of DNA without describing its chemistry would be superficial and incomplete.

Molecules, a group of atoms bonded together, come in different sizes. When a molecule is very large, more than a thousand atoms, it is referred to as a macromolecule. Even very large macromolecules, such as DNA with millions of atoms, can only be seen with an electron microscope, not a light microscope. All naturally occurring macromolecules, such as proteins and nucleic acids, are constructed from a limited number of building blocks or subunits.

It's in Your DNA. http://dx.doi.org/10.1016/B978-0-12-812502-1.00002-0

Determination of the structure of a macromolecule involves the following steps: first, the macromolecule has to be isolated and purified so that it contains little or no contaminating materials. Second, the chemical structure of each of the component parts, or building blocks, of the macromolecule must be determined. Third, it is necessary to clarify how the building blocks are joined together to form the macromolecule. Finally, the three-dimensional structure of the molecule must be resolved. This process is reminiscent of an archeological excavation: first one finds the little pieces, then one tries to understand what they are, and finally one builds from them entire three-dimensional structures.

Initial research on the chemical structure of nucleic acid was carried out in the early 1900s by the biochemist Phoebus Levene, working at the Rockefeller Institute, New York. Levene was born in Russia in 1869 and studied medicine at the Imperial Military Medical Academy in St. Petersburg. As a student, he worked in the laboratory of his chemistry professor where he likely developed an interest in biochemistry. In 1891, because of growing anti-Semitism in Russia, Levene and his family immigrated to the United States. Shortly after arriving in New York City, he began practicing medicine on the Lower East Side. However, Levene did not give up research. He enrolled as a special student at Columbia University and he split his time between his medical practice and research in the department of physiology. By 1894, he began publishing papers on the chemical structure of sugars. Two years later, Levene received his first appointment as an Associate in the Pathological Institute of the New York State Hospitals. Unfortunately, around this time, Levene contracted tuberculosis and was forced to take time off to recuperate. Levene used the time between 1896 and 1905 to regain his health and to work with a number of well-known chemists, including Albrecht Kossel in Germany, the nucleic acid expert of the time. In 1905, Levene was hired by the newly established Rockefeller Institute of Medical Research to head the biochemical laboratory. Levene did most of his nucleic acid work at the Rockefeller and stayed there until his death in 1940.

Levene was a cultured man, an art lover and a collector. The walls of his house were lined with prints and paintings or overflowing bookshelves. Levene was extremely well-read and was fluent in Russian, English, French, and German. He also spoke passable Spanish and Italian. His experience, knowledge, and his generosity made him a favorite with colleagues and friends. He was also said to be a great teacher, enthusiastic, and supportive.

Levene's research began with the isolation and purification of nucleic acid from calf thymus, which was readily available in New York butcher shops and an excellent source of nucleic acid. (The thymus is an organ that is located in the upper chest behind the breastbone and in front of the lower neck in which the immune cells mature and multiply.) The procedure he developed in order to purify nucleic acid consisted of dissolving trimmed and finely ground thymus gland in boiling water containing caustic soda, acidifying the solution, and then adding the common alcohol, ethanol. On standing overnight, the nucleic acid settles as a spongy white mass which can be separated from the fluid. He then washed the nucleic acid with ethanol by decantation several times. The product, referred to as thymus nucleic acid, can scarcely be improved by any further purification. The same method can be applied to other animal tissues. Although considerably purer than Miescher's nuclein that I described in the previous chapter, thymus nucleic acid contained similar properties. (Today, one can purchase calf thymus and salmon sperm nucleic acid and many other kinds of DNA from biochemical supply companies; alternatively, nucleic acid can be isolated from a wide variety of sources using commercially available nucleic acid isolation kits.)

Once Levene isolated calf thymus nucleic acid in a pure form, his next task was an analysis of its component parts. The nucleic acid was suspended in a solution of strong acid and then heated to break the molecule into a mixture of small parts. The small pieces were then separated according to their chemical properties. Each component of thymus nucleic acid was identified by comparing its properties, such as its melting point (the temperature at which a solid becomes a liquid) and its precise elementary composition of carbon (C), hydrogen (H), oxygen (O), and nitrogen (N) to those of known compounds.

After almost 30 years of hard work, Levene and other scientists identified each of the six building blocks or subunits of nucleic acid. Thymus nucleic acid was composed of one part sugar, one part phosphoric acid, and approximately one-fourth part of each of the nitrogen-bearing ring compounds, referred to as nitrogenous bases. When the sugar was shown to be deoxyribose, thymus nucleic acid was referred to as deoxyribonucleic acid or DNA for short (Figs. 2.1 and 2.2).

The four nitrogenous bases of DNA, eventually shown to play an important role in cellular information transfer, are: thymine (T), cytosine (C), adenine (A), and guanine (G). Where did these names come from? The names of chemical molecules are often chosen according to the source from where the molecule was first isolated. Guanine was initially discovered in guano, the waste product of birds and bats and a favorite fertilizer of organic farmers. Thymine was first isolated from the thymus gland, and adenine was initially extracted from the pancreas (a specific gland; in Greek, "*aden*"). The name cytosine was derived from cyto- "cell" + chemical suffix -ine. The chemical structures of the four bases of DNA are shown earlier. Thymine and cytosine, referred to as pyrimidines, are molecular rings formed of four carbon and two nitrogen atoms. Adenine and guanine, referred to as purines, are composed of two fused linked rings containing five carbon and four nitrogen atoms. The fact that the purines (A and G) are slightly larger than the pyrimidines (T and C) will become relevant when we consider the three-dimensional structure of DNA.

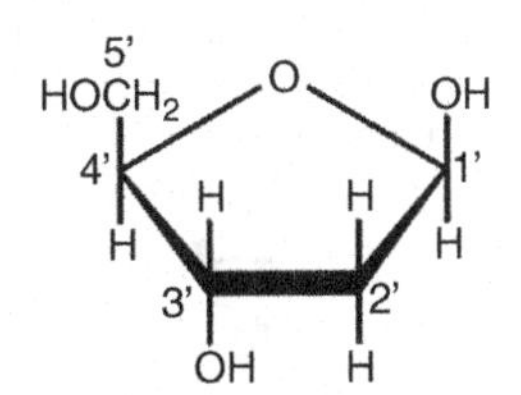

FIGURE 2.1 2-Deoxyribose.

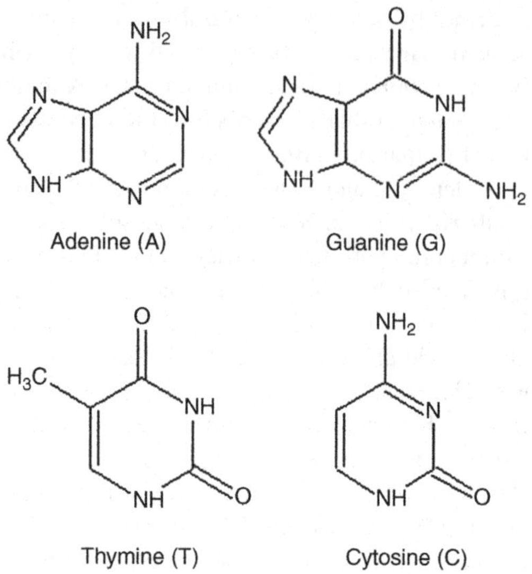

FIGURE 2.2 The four bases found in DNA. A and G are double-ring purine bases. T and C are single-ring pyrimidine bases.

Biological macromolecules are generally polymers constructed from monomers, proteins are polymers built from amino acid monomers, for example, and polysaccharides, such as starch and cellulose, are made from sugar monomers. We know now that an intact DNA molecule is a giant polymer of great complexity. Even the simplest virus DNA molecule contains more than 5000 subunits, each of which is built from one molecule of the sugar—deoxyribose, one molecule of phosphoric acid, and one molecule of one of the four nitrogenous compounds—adenine, guanine, cytosine, or thymine. The subunit of DNA, consisting of the sugar, phosphorus, and one of the bases is referred to as a nucleotide. Large numbers of nucleotide subunits in DNA are required to store genetic information (Fig. 2.3).

The next question that Levene set out to solve was: how are the four nucleotides joined together to form the giant DNA molecule? Levene showed that the components of DNA were joined together to form a long polymeric chain of alternating sugar (deoxyribose) and phosphoric acid units, with one of the four bases, adenine, guanine, cytosine, or guanine, joined to each deoxyribose (Fig. 2.4).

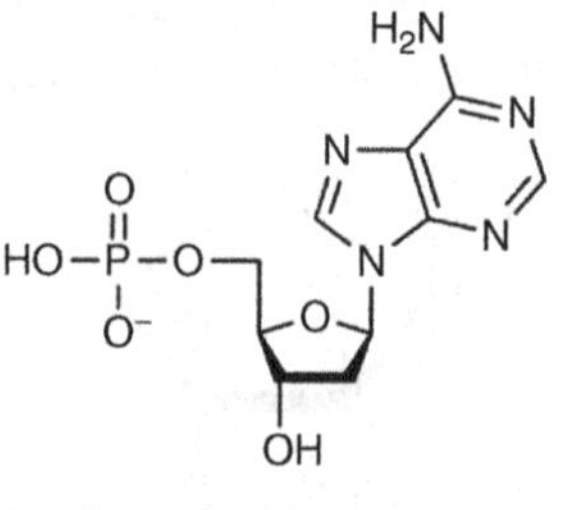

FIGURE 2.3 Chemical structure of a nucleotide, the subunit of DNA. This nucleotide contains the deoxyribose, a nitrogenous base called adenine, and one phosphate group.

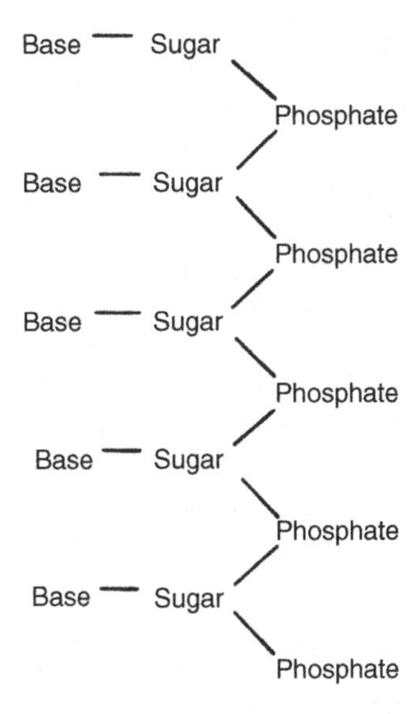

FIGURE 2.4 The two-dimensional structure of DNA. The backbone consists of a long chain of deoxyribose sugars joined by phosphate groups. Each sugar contains one of the four bases, thymine, cytosine, guanine, or adenine.

As mentioned earlier, most of the early research on DNA chemistry was carried out with either salmon sperm DNA or calf thymus DNA. It so happens that both of these substances have approximately equal amounts of the four bases, adenine, guanine, cytosine, and thymine. As I shall show later in this chapter, the presence of equal amounts of the four bases is not generally true. However, Levene was unaware of these differences and based his hypothesis of DNA structure on the equivalence of the four bases in salmon and calf thymus DNA. The simplest DNA structure to explain the facts known to Levene is what he called the "repeating tetranucleotide hypothesis." According to this hypothesis, which was widely accepted until the 1940s, the four bases would repeat in the same order throughout the length of the molecule. It was also accepted that such a regular structure could not carry enough genetic information to serve as a basis of heredity. Instead, the protein component of chromosomes was thought to be the basis of heredity, and therefore before the 1940s most research on the physical nature of the genes focused on proteins and particularly on enzymes.

To be useful, a scientific hypothesis, such as the "repeating tetranucleotide hypothesis," should be experimentally testable. Disproving a hypothesis often contributes more to scientific progress than providing support for one. Experiments that are consistent with a hypothesis cannot prove that a hypothesis is correct because there may be other hypotheses that are consistent with the facts, but if a well performed and repeatable experiment is not consistent with the hypothesis, then the hypothesis must be discarded. This was the case with the tetranucleotide hypothesis. The biochemist Erwin Chargaff disproved the tetranucleotide hypothesis by demonstrating that DNA isolated from different organisms contained different ratios of A, G, C, and T. In fact, Chargaff's analysis of the ratios of the four bases provided information that would eventually help understand how DNA could store genetic information and be able to replicate itself.

Chargaff was born in 1905, in Chernivtsi, then part of the Austro-Hungarian Empire. At the outbreak of World War I, his family moved to Vienna, where he attended the Gymnasium Wasagasse and subsequently the Vienna University of Technology where he received a doctorate in chemistry in 1928. After graduation he became an assistant at the University of Berlin. Chargaff was Jewish, and Nazi policies put in place when Hitler came to power in 1933, excluded Jews from academic positions. As a result, he immigrated to the United States, and started his lifelong career at Columbia University. Hans Thacher Clarke, the Chairman of the Biochemistry Department at Columbia University, graciously opened his Department to several refugee biochemists, among them was Erwin Chargaff.

In order to examine Levene's tetranucleotide hypothesis, Chargaff needed a technique for separating and accurately determining the relative amounts of A, G, C, and T in a variety of DNA samples from different tissues and organisms. Fortunately, the powerful technique of paper chromatography for separating small molecules was developed in 1943 by the British chemists Arthur Martin and Richard Synge. For their discovery of paper chromatography, Martin and Syrnge received the Nobel Prize in Chemistry in 1957.

Paper chromatography is in practice as an exceedingly simple procedure. A small drop of the material to be examined is placed about 2 in. (5 cm) from the edge of a rectangular sheet of filter paper and allowed to dry there. The edge of the paper is then dipped in an appropriate liquid (the solvent) that moves up the paper by capillary action. As the liquid passes over the point of application each chemical compound is pulled along by the liquid at a speed which is characteristic of that substance. By using this technique, a mixture of substances applied to the same spot can be separated from each other on the filter paper in a few hours because they move with the liquid at different speeds. For example, if a mixture of A, G, C, and T is applied on one spot, and the liquid solvent butyl alcohol is allowed to pass over it, the four substances will separate from each other in approximately 18 h (see later). Thymine moves the fastest, followed by cytosine, adenine, and guanine (Fig. 2.5).

In the 1980s, paper chromatography was replaced by more rapid and efficient technique of HPLC (high performance liquid chromatography). HPLC works on the same principle as paper chromatography, but high pressure is used to generate the liquid flow. HPLC is now one of the most powerful tools in analytical chemistry. It has the ability to separate, identify, and quantitate the compounds that are present in any sample that can be dissolved in a liquid. Today, compounds in trace concentrations as low as parts per trillion may easily be identified. HPLC can be, and has been, applied to just about any sample, such as pharmaceuticals, food, nutraceuticals, cosmetics, environmental matrices, forensic samples, and industrial chemicals.

FIGURE 2.5 Separation of DNA bases by paper chromatography, as observed by Chargaff.

The precise amount of each component can be determined by how much UV light is absorbed. The fact that the four bases in DNA strongly absorb UV light is the reason why exposure to sunlight causes mutations in the DNA which may result in skin cancer. Although most of the UV light energy absorbed by DNA is released as heat, a significant fraction of the energy results in chemical damage to the DNA. When the cell replicates the damaged DNA, errors are produced that can lead to mutation and cancer (discussed in Chapter 16).

Using the simple technique of paper chromatography, Chargaff and his collaborators analyzed the base composition of DNA from various sources. By the early 1950s, enough information had accumulated to draw some very interesting conclusion (Table 2.1).

Conclusion one. Levene's tetranucleotide hypothesis is not correct because the hypothesis predicts that all four bases are in equal abundance in all organisms.

Conclusion two. The base composition of DNA is characteristic of the species, differing in composition for different species, but not for tissues of any one species.

Conclusion three. For all DNA samples the amount of adenine equals the amount of thymine (A = T) and guanine equals cytosine (G = C).

As Chargaff wrote, "The results serve to disprove the tetranucleotide hypothesis. It is, however, noteworthy—whether this is more than accidental cannot yet be said—that in all deoxyribonucleic acids examined thus far the molar ratios of total purines to total pyrimidines, and also of adenine to thymine and of guanine to cytosine, were not far from 1."

TABLE 2.1 Average Base Composition of DNA of Several Species[a]

Source of DNA	Adenine (A)	Guanine (G)	Cytosine (C)	Thymine (T)
Humans	29.9	20.7	20.3	29.1
Calf spleen	29.6	20.4	20.8	29.2
Calf thyroid	29.5	20.9	20.8	29.0
Calf thymus	28.8	21.4	21.7	28.1
Yeast	31.7	18.1	17.6	32.6
Micrococcus (bacteria)	14.5	36.1	35.9	13.5
Salmon sperm	28.7	21.8	21.4	28.1
Rickettsia (bacteria)	33.8	16.7	15.9	33.6
Cotton plant	32.7	17.1	17.4	32.8
Plasmodium (protozoa)	40.2	10.1	9.9	39.8

[a]These results are expressed as percentages of each base in the DNA. The data are accurate to ± 1%. In animal and plant cells, the base composition can vary from one chromosome to another. Average values are presented.

Although these conclusions, especially the third one, were important in the subsequent determination of the double helix structure of DNA, Chargaff was not included in the group that eventually received the Nobel Prize for the discovery of the double helix structure of DNA. I assume that this was partly because he was an abrasive man having difficulty with empathy and social skills. (On a personal note, Chargaff served on the examining committee for the oral defense of my PhD thesis at Columbia University in 1961. He ended the exam by looking at me and uttering through the pipe in his mouth, "I think we have squeezed enough blood out of this turnip.")

NOTES AND REFERENCES

10 Tipson, R.S., 1957. Phoebus Aaron Theodor Levene, 1869–1940. Adv. Carbohydr. Chem. 12, 1–12.

13 Hargittai, I., 2009. The tetranucleotide hypothesis: a centennial. Struct. Chem. 7, 753–756.

14 Martin, A.P., 1952. The development of partition chromatography. Nobel Lecture, December 12, 1952. Nobel Lectures, Chemistry 1942–1962. Elsevier Publishing Company, Amsterdam.

15 Chargaff, E., Zamenhof, S., Green, C., 1950. Composition of human desoxypentose nucleic acid. Nature 165, 756–757.

15 Rünger, T.M., Kappes, U.P., 2008. Mechanisms of mutation formation with long-wave ultraviolet light (UVA). Photodermatol. Photoimmunol. Photomed. 24, 2–10.

16 Chargaff, E., 1978. Heraclitean Fire–Sketches From a Life Before Nature. Rockefeller University Press, New York, NY.

16 Wright, P., 2002. Erwin Chargaff: disillusioned biochemist who pioneered our understanding of DNA. Obituaries in The Guardian.

Chapter 3

DNA Is the Genetic Material

Philosophy begins when you ask a general question and so does science.

—Bertrand Russell

There should be a science of discontent. People need hard times and oppression to develop psychic muscles.

—Frank Herbert author of *Dune*

Before we learn about the crucial experiments demonstrating that DNA is the genetic material, we should understand the development of the gene concept. The science of genetics can be traced back to the carefully controlled studies of the Austrian monk, Gregor Mendel, around the mid-19th century.

Born in 1822 on his family's farm in Heinzendorf, Austria (now Hynčice, Czech Republic), Johann Mendel was raised in a rural setting. His academic abilities were recognized by the local schoolmaster, who persuaded his parents to send him away to school at the age of 11. His Gymnasium studies completed with honors in 1840, Mendel entered a 2-year program in philosophy at the Philosophical Institute of the University of Olmütz, Czech Republic, where he excelled in physics and mathematics, completing his studies in 1843. His initial years away from home were hard because his family could not sufficiently support him. He tutored other students to make ends meet, and twice he suffered serious depression and had to return home to recover. As his father's only son, Mendel was expected to take over the small family farm, but he preferred a different solution to his predicament, choosing to enter a monastery of the Augustinian order, where he was given the name Gregor.

The move to the monastery took him to Brünn, the capital of Moravia, where for the first time he was freed from the harsh struggle of former years. He was also introduced to a diverse and intellectual community. As a priest, Mendel found his parish duty to visit the sick and dying so distressing that he again became ill. The monastery found him a substitute-teaching position at Znaim (Znojmo, Czech Republic), where he proved very successful—and was sent to the University of Vienna for 2 years to benefit from a new program of scientific instruction. Mendel devoted his time at Vienna to physics and mathematics. He also studied the anatomy and physiology of plants and the use of the microscope. In the summer of 1853, Mendel returned to the

It's in Your DNA. http://dx.doi.org/10.1016/B978-0-12-812502-1.00003-2

FIGURE 3.1 Johann Gregor Mendel (1822–84).

monastery in Brünn, and in the following year he was again given a teaching position, this time at the Brünn Realschule (secondary school), where he remained until elected as head of the monastery 14 years later. These years were his greatest in terms of success both as teacher and as skillful experimentalist (Fig. 3.1).

Mendel chose to research the transmission of hereditary traits in common pea plants for a number of reasons. First of all, the common pea exhibits many distinct varieties. Second, they can be grown easily in large numbers. Third, their reproduction can be manipulated and offspring could be quickly and easily produced. Pea plants have both male and female reproductive organs. As a result, they can either self-pollinate or cross-pollinate with another pea plant. In his experiments, Mendel was able to selectively cross-pollinate plants with particular traits and observe the offspring over many generations. He conducted his experiments in the Augustinian monastery's 4.9 acre experimental garden in old Brno, Czech Republic. He cross-fertilized pea plants with clearly different characteristics—tall with short, purple flower color with white flower color, those containing green pods with those containing yellow pods, etc.

At the time of Mendel's studies, it was a generally accepted that the hereditary traits of the offspring of any species were merely the diluted blending of whatever traits were present in the "parents." However, the results of such studies were often skewed by the relatively short period of time during which the experiments were conducted, whereas Mendel's research continued over as many as 8 years (between 1855 and 1863), and involved tens of thousands of individual plants. Analyzing his results, Mendel was able to determine that traits are passed from parents to offspring with mathematical precision in small, separate information packets that subsequently become known as genes (from the Greek *genea* "generation, race"). The Danish botanist Wilhelm Johannsen first coined the word "gene" in 1909.

Consider the following experiment performed by Mendel: first, a pure bred plant with green pods was cross-fertilized with one with yellow pods. All of the progeny of these crosses had green pods. He then allowed all of the progeny from the first cross to self-pollinate. From the self-pollination experiment, Mendel noticed a 3:1 ratio in pod color: about three-fourth of the plants had green pods and about one-fourth had yellow pods. From these experiments Mendel formulated what is now known as Mendel's Law of Segregation, which established that there are dominant and recessive traits passed on randomly from parent to offspring. In the earlier example, one parent had two genes for green pods and the other parent had two genes for yellow pods. Thus, all of the resulting offspring had one gene for green pods and one gene for yellow pods. Since these hybrid offspring all had green pods, green must be dominant and yellow recessive. In the self-pollination experiment, random assortment of the genes from the hybrid plants would yield one-fourth of the plants with two genes for yellow pods, one-fourth of the plants with two genes for green pods, and half of the plants with one gene for green pods and one gene for yellow pods. Since green is dominant over yellow, the resulting ratio in the color, green to yellow, of the offspring pods was 3:1. These types of experiments were performed for many generations using a variety of different traits. They established not only the Law of Segregation but also the Law of Independent Assortment, which established that traits were passed on independently of other traits from parent to offspring.

In 1865, Mendel delivered two lectures on his findings to the Natural Science Society in Brno, and the results of his studies were published the following year, under the title "Versuche über Pflanzenhybriden" (Attempts at Plant Hybridization). However, his breakthrough findings were largely misunderstood and ignored for 35 years. Two factors contributed to the lack of appreciation of Mendel's breakthrough discoveries: first, the findings were in conflict with the existing dogma. Second, Mendel did little to promote his work. Gregor Mendel died on January 6, 1884, at the age of 62. He was laid to rest in the monastery's burial plot, and his funeral was well attended. His work, however, was still largely unknown.

In 1900, a team of botanists independently duplicated Mendel's experiments and results, finding out after the fact, allegedly, that both the data and the general theory had been published in 1866 by Mendel. After a short time, the scientists did credit him with priority and Mendel is rightly considered the "father of modern genetics." The importance of Mendel's theory of inheritance was then recognized and widely accepted, and became the starting point for further investigations on the nature of inheritance that were carried out by Thomas Hunt Morgan and other 20th-century geneticists. Mendelism, as the theory was called, was merged with Darwinism in the 1930s to form the "New Synthesis," which explained evolutionary theory in modern genetic terms.

Shortly after the rediscovery of Mendel's work in 1900, a group of scientists at Columbia University, led by Thomas Hunt Morgan began an extensive study of the genetics of a small fruit fly, *Drosophila melanogaster*. Morgan, born in

Kentucky in 1866, came from a distinguished southern family. His uncle, Charles Hunt Morgan, was a Confederate General and the great-grandson of Francis Scott Key, the composer of the "Star Spangled Banner." Morgan received his PhD from Johns Hopkins University in zoology in 1890 for his research on the embryology of sea spiders. Following the rediscovery of Mendelian inheritance in 1900, Morgan's research moved to the study of mutation in the fruit fly *D. melanogaster*. The research was performed in what became known as the "Fly Room" at Columbia University.

One of Morgan's major contributions, a sociological one, was the introduction at Columbia University and into American science as a whole a set of sweeping institutional changes. Until the start of the twentieth century, the leading American research universities, Harvard, Johns Hopkins, Columbia, and Chicago, had all been inspired by the model of the German research university, in which the *Geheimrat,* the great scientific leader, ordered the hierarchy of his subordinates. Morgan, however, based laboratory governance on democratic principles of merit rather than seniority. If one were to ask scientists around the world what is unique about America, they point to the university, and to this day foreign scientists are amazed that students working in a laboratory call professors by their first names. Morgan surrounded himself with a brilliant group of undergraduate and graduate students. He encouraged the free exchange of ideas in an atmosphere that was at once friendly, yet self-critical.

The atmosphere in the Fly Room was described by one of his outstanding students, Alfred Henry Sturtevant:

> *This group worked as a unit. Each carried on his own experiments, but each knew exactly what the others were doing, and each new result was freely discussed. There was little attention paid to priority or to the source of new ideas or new interpretations. What mattered was to get ahead with the work. There was much to be done; there were many new ideas to be tested, and many new experimental techniques to be developed. There can have been few times and places in scientific laboratories with such an atmosphere of excitement and with such a record of sustained enthusiasm. This was due in part to Morgan's own attitude, compounded with enthusiasm combined with a strong critical sense, generosity, open-mindedness, and a remarkable sense of humor.*

The major advantages of the fruit fly compared to the pea plant are shorter generation time, larger progeny, and the research could be conducted in the laboratory under controlled conditions. Shortly after Morgan began work with the fruit flies, he observed that a number of spontaneous mutations appeared in the fly cultures, including white eyes, cut wings, curly wings, yellow body color, and brown eyes. Using these mutants, Morgan showed that the genes responsible for these mutant traits were located in a linear arrangement on chromosomes. Chromosomes are threadlike structures containing DNA and protein which are present in the nucleus of all animals and plant cells. The word chromosome comes from the Greek *χρῶμα* (chroma, "color") and *σῶμα* (soma, "body").

Chromosomes are strongly stained by particular dyes. For his pioneering genetic studies on fruit flies, Thomas Morgan received the Nobel Prize in 1933. In typical Morgan generosity, he shared the prize money with two of his former students, Calvin Bridges and Alfred Sturtevant. Although Morgan officially retired in 1941, he continued to work in the laboratory until his death in 1945.

The concept of a gene was invented in order to explain certain patterns of inheritance. [The Danish botanist Wilhelm Johannsen coined the word "gene" ("*gen*" in Danish and German) in 1909 to describe this fundamental physical and functional unit of heredity]. The genes were characterized as "information packets" that are located at particular points on particular chromosomes. They constitute the heredity organelles of the cell. In effect, the genes were treated as though they were strings of beads of unknown composition that are duplicated and transmitted from parent to offspring by an unknown mechanism and that somehow interact with the environment to determine the characteristics of the organism. The immediate and most fundamental question to ask at that point was: what is the chemical nature of the gene? The answer would come from experiments with bacteria and viruses.

In 1928, Fredrick Griffith, a medical officer in the British Ministry of Health, reported the results of his studies on what at that time was the number one killer of humans, pneumonia. It was known that pneumonia was caused by a spherically shaped bacterium, named pneumococcus. Griffith began his studies by isolating the bacterium from the sputum of patients with pneumonia. When the pneumococcus was initially isolated, it formed only one type of colony on the agar medium, later called type S, which had a mucoid or smooth (S) appearance. When the type S bacteria were injected into mice, the mice contracted pneumonia and died. When type S bacteria were transferred from one agar plate to another, there occasionally appeared (we now know, by mutation) a second type of colony that appeared rough, referred to as type R. This type was unable to cause the disease when injected into mice.

Griffith realized that the basic difference between type S and type R is that type S contained a capsule surrounding the cell. These capsules help produce the smooth appearance of the colony, and more importantly, protect the bacteria so they can survive and grow in the animal body. Type R, lacking the capsule, gives rise to rough colonies and cannot survive in the animal because it does not have the capsule.

To understand how these capsules participated in the infectious process, Griffith performed a series of experiments on mice. Keep in mind that the motivation for these experiments was not a curiosity regarding the nature of the gene, but rather a quest for information about the mechanism of the infection process. As so often happens in science, the correct interpretation of these experiments came much later. For the moment we shall consider only Griffith's results, summarized below: the interpretation and significance will emerge as subsequent experiments are discussed (Table 3.1).

TABLE 3.1 Griffith's Original Transformation Experiments

Experiment	Injection into mice	Result
1	Type R only[a]	No effect
2	Type S only[b]	All mice dead
3	Heat-killed type S[c]	No effect
4	Type R plus heat-killed type S	All mice dead

[a] *Type R: no capsule.*
[b] *Type S: capsule.*
[c] *The bacterial culture was heated to 65°C.*

Experiments 1 and 3 were control experiments, showing that neither type R nor heat-killed type S was able to cause the disease when injected separately into mice. However, when they were mixed together and injected, the mixture was lethal. Furthermore, a postmortem examination of the mice from experiment 4 revealed only live type S pneumococci. When the smooth colonies were transferred to fresh agar media, they bred true; that is, the smooth characteristic was retained and reproduced in subsequent generations.

The process by which type R were converted to type S came to be known as *transformation*. As we will discuss in the next section, transformation involves the replacement of one gene by another. Although Griffith performed careful experiments and the process of transformation was confirmed by other workers, it did not attract due attention. One possible reason for the apparent lack of interest in Griffith's experiments was that he did not emphasize the genetic implications of his work. In 1941, Griffith was killed during World War II's London Blitz and did not live to see the significance of his research appreciated.

Continuing the research done by Griffith, Oswald T. Avery and his team at the Rockefeller Institute discovered that transformation of type R pneumococci into type S could be carried out in the absence of mice. When they mixed live type R bacteria with heat-killed type S and placed the mixture on an appropriate agar medium, approximately 1% of the colonies that appeared were smooth (type S). To make sure the formation of type S bacteria was the result of transformation, Avery performed two control experiments. In the first control, he showed that the sample of heat-killed type S bacteria gave no colonies when placed on the medium, indicating that the heat treatment was effective in killing all the bacteria. In the second control, he showed that plating type R alone yielded only 0.00001% smooth colonies, a typical low value for spontaneous mutation. Since the frequency of smooth colonies arising from the mixture was much higher than the controls, Avery concluded that the process of transformation, or gene replacement, must have taken place.

It is reasonable to ask at this time: why, when the mixture of type R and heat-killed type S was injected into a mouse, were *all* of the bacteria found at the postmortem examination type S, where when placed directly on the agar medium, only 1% were type S? The answer is that in all probability, Griffith also obtained only 1% transformation in his cultures, but that when injected into mice, the type R was destroyed (probably by the mouse immune system) and did not show up in the postmortem examination.

Avery concluded from his experiments and those of Griffith that heat-killed type S bacteria contain an active substance that enters into the live type R cells and transforms them into type S. Since this trait is inherited in subsequent generations, it has the properties that are uniquely associated with the gene. Avery referred to this active substance as *transforming principle (= gene)*. The Rockefeller group then began an extensive program to isolate and identify the transforming principle.

Biochemists have developed, mostly by trial and error, a series of techniques for isolating and identifying chemical substances from cellular matter. The key steps in the isolation of the transforming principle from pneumococcus were breaking the cells with a detergent and precipitating the DNA with alcohol. After the pneumococcus type S DNA was collected and dissolved in water, it was added to type R cells and the mixture placed on agar medium. Approximately 1% of the colonies arising on the medium were type S, indicating that the DNA preparation contained the transforming principle. However, it was still possible that small amounts of contaminating materials, such as proteins, were responsible for the transformation. To eliminate this possibility, Avery and his coworkers treated the DNA preparation with different kinds of enzymes and again tested for transformation. The only enzyme that caused the loss of the transforming activity was the enzyme named DNase that specifically destroys DNA. Treatment with proteinases (enzymes that destroy proteins) had no effect on the activity. Thus, the conclusion was that *the gene must be made up of DNA* (at least in this case).

Avery was 67 when his extraordinary results were first reported in 1944, demonstrating that not all breakthrough discoveries are achieved by young scientists. Nobel laureate Joshua Lederberg stated that Avery and his laboratory provided "the historical platform of modern DNA research and betokened the molecular revolution in genetics and biomedical science generally." The Nobel laureate Arne Tiselius said that Avery was the most deserving scientist to not receive the Nobel Prize for his work, though he was nominated for the award several times. It should me mentioned that not all great works are appreciated at the time of their discovery (Fig. 3.2).

Shortly after the results of the Rockefeller group were published, other scientists showed that: (1) a large number of other traits could be transformed. For example, DNA extracted from strains of pneumococci that were resistant to the antibiotic streptomycin could transform streptomycin-sensitive strains into the resistant form. (2) DNA transformation occurred in other bacteria, such as *Bacillus subtilis* and *Haemophilus influenza*. By 1950, the evidence from these and other experiments was convincing that in bacteria, at least, the genetic material is DNA.

FIGURE 3.2 **Owald T. Avery (1877–1955).**

In 1952, the microbiologist Alfred Day Hershey with his research assistant, Martha Chase, at the Cold Spring Harbor laboratories in New York performed a simple but elegant experiment demonstrating that DNA is also the genetic material of viruses. The logic behind their experiment, referred to as the "blender experiment," was based on two accepted properties of viruses: (1) most viruses are composed of only two types of molecules, proteins and nucleic acids. (2) When viruses infect cells, only one part of the virus enters the cells. Using this information, Hershey and Chase used viruses to test which part of the virus entered the cell, protein or DNA? One batch of viruses was made radioactive with phosphorus atoms (present in DNA, but not protein) and another with radioactive sulfur atoms (present in protein, but not DNA). They then used the two batches to infect separate cultures of fresh bacteria. After allowing just enough time for the bacterial viruses to infect the bacteria cells, the mixtures were shaken vigorously to dislodge parts of the phage that became attached to the outside of the bacteria but did not actually penetrate them (a common kitchen blender was used). Accordingly, only that part of the phage that entered the bacterium remained with it during purification. The purified bacteria were then examined for radioactivity. The result was that only bacteria that were infected with phages that contained radioactive phosphorus were radioactive. Therefore, the only part of the virus that enters the bacteria and must therefore contain the genetic information is DNA (Fig. 3.3).

Today we know that not in all viruses contain DNA as the genetic material. Some viruses contain RNA (ribonucleic acid) instead of DNA as their genetic material. Notable human diseases caused by RNA viruses include SARS, influenza, hepatitis C, West Nile fever, polio, and measles.

Hershey became director of the Carnegie Institution in 1962 and was awarded the Nobel Prize in Physiology or Medicine in 1969, shared with Salvador Luria and Max Delbrück for their discovery on the replication of viruses and their genetic structure. In 1974, Hershey retired, though he was still a regular visitor

FIGURE 3.3 **DNA of bacterial viruses entering a bacterium.**

to Cold Spring Harbor Laboratory. In 1979, a building on the grounds was dedicated to him. Hershey was known to be an excellent writer and editor. His papers were clear and concise and he helped young scientists learn the craft of scientific writing. He enjoyed gardening and woodworking, as well as classical music. In the early 1980s, he became interested in computers and used them to catalog his classical music collection. He was busy, active, and still learning until his death in 1998 at the age of 89.

In plants and animals, including humans, a large body of evidence indicates that here too the genetic material is DNA. To begin with, all cells contain DNA that is localized on chromosomes in the nucleus of the cell. This is consistent with the fact that the genes are also located specifically on chromosomes. Second, within the same species, the amount of DNA per cell is the same in all organs. Haploid cells, such as spermatozoa and unfertilized eggs, which contain only one set of chromosomes, that is, one half of the amount of genetic material), contain also only one half the amount of DNA. Most convincing are the more recent genetic engineering experiments (discussed in Chapter 10) in which pieces of DNA are introduced into animal and plant cells and genetically change them.

Once it became clear that DNA was the genetic material, the next fundamental question was: how does it work? How does DNA replicate itself so that genetic information is accurately transmitted from parent to offspring, and how is genetic information, stored in DNA, translated by living organisms into observable traits? For example, how does a gene for blue eyes actually cause an individual to exhibit that pigmentation? The answers to these questions followed the discovery of the three-dimensional structure of DNA, the double helix.

NOTES AND REFERENCES

17 Bowler, P.J., 2003. Evolution: The History of an Idea. University of California Press, Berkeley.

19 Bateson, W., 2009. Mendel's Principles of Heredity: A Defence, with a Translation of Mendel's Original Papers on Hybridization (Cambridge Library Collection—Life Sciences). Cambridge University Press, Cambridge, UK.

19 Weinstein, A., 1977. How unknown was Mendel's paper? J. Hist. Biol. 10, 341–364.

20 Fisher, R.A., De Beer, G.R., 1947. Thomas Hunt Morgan. 1866–1945. Obituary Notices Fellows Roy. Soc. 5 (15), 451–466.

20 Lobo, I., Shaw, K., 2008. Thomas Hunt Morgan, genetic recombination, and gene mapping. Nat. Edu. 1, 205.

22 Downie, A.W., 1972. Fourth Griffith Memorial Lecture. Pneumococcal transformation—a backward view. J. Gen. Microbiol. 73 (1), 1–11.

23 Russell, N., 1988. Oswald Avery and the origin of molecular biology. Br. J. Hist. Sci. 21, 193–400.

23 Hotchkiss, R. D., Weiss, E., 1956. Transformed bacteria. Scientific American, November 1956. This is a popular account of the discovery and early development of transformation.

24 Campbell, A., Stahl, F.W., 1998. Alfred D. Hershey. Annu. Rev. Genet. 32, 1–6.

Chapter 4

DNA in Three Dimensions: The Double-Helix

"Faith" is a fine invention
For Gentlemen who see!
But Microscopes are prudent
In an Emergency!

—Emily Dickinson

When Winston Churchill was asked how history will look upon him, he replied, "history will look favorably upon me, because I will write the history." The same is true regarding the history of the discovery of the double helix structure of DNA, written by one of its discoverer's, James Watson. His book, *The Double Helix*, written in 1968, 15 years after the discovery, is a personal account of the scientific and social interactions that led to this immensely important discovery. The book has been praised by many for providing a readable account of the beautiful experience of making a great scientific discovery and condemned by some for being too subjective, in addition to criticizing and not giving credit to other scientists who contributed to the discovery.

After reading Watson's manuscript, the codiscoverer of the DNA double-helix structure, Francis Crick, urged Watson not to publish the book: "the book is not a history of the discovery of DNA, as you claim in the preface. Instead it is a fragment of your autobiography. Anything which concerns you and your reactions, apparently, is historically relevant, and anything else is thought not to matter. If you publish your book now, in the teeth of my opposition, history will condemn you." To a large extent, Crick's prediction has been proven correct.

Objectivity was partly restored when in 1975 Anne Sayre published a highly informative biography of Rosalind Franklin, who was a central figure in the discovery of the double helix structure of DNA, documenting how Watson had distorted the truth and how Franklin deserved considerably greater credit than she was given. The editor of the *Journal of the American Association for the Advancement of Science* reviewed Sayre's book and wrote "as the true story of a female scientist's efforts in the field of biochemistry, this book should be required reading for all aspiring scientists—especially women."

It's in Your DNA. http://dx.doi.org/10.1016/B978-0-12-812502-1.00004-4

Anne Sayre points out in her biography of Rosalind Franklin that Watson misrepresents Franklin both as a person and a scientist. To begin with, Watson refers to Franklin as "Rosy," a nickname that was not used by any of her friends. He characterizes "Rosy" as unattractive, aggressive, overbearing, "unfeminine" know-it-all, the female stereotype we have been taught to fear or despise. According to Watson, "Rosy" peers out at the world from behind her spectacles. According to Sayre, who knew Franklin well, this description of Franklin is absurd. Franklin's friends and the scientists she worked with portrayed her as strikingly good looking, always well-dressed, smart and strong, with a reserved nature. The fact that she had perfect eyesight and never wore glasses makes one wonder at the overall truthfulness of Watson's account of the discovery of the DNA structure.

Why did Watson make up the character of "Rosy?" Besides his obvious belief that women do not belong in science, portraying Franklin as a rigid person was meant to detract from her major contribution. Introducing Franklin, Watson writes, "The real problem was 'Rosy.' The thought could not be avoided that the best home for a feminist is in another person's lab." Watson claimed that Franklin was working in a laboratory under the direction of Maurice Wilkins, whereas she was totally independent, working in her own laboratory. She was the expert on X-ray techniques, not Wilkins.

Before describing the brilliance of the Watson–Crick–Franklin discovery, I will first explain the crucial experimental evidence upon which it was based. To begin with, as we remember from the previous chapter, the work of Avery, Hershey, and Chase and others showed that DNA was the genetic material. Watson, Crick, and Franklin were aware of this and each one of them grasped the importance of elucidating the 3D structure of DNA in order to understand how genes function. As discussed in Chapter 2, the two-dimensional chemical structure of DNA was known from the work of Levene and others. It consisted of a long chain of deoxyribose molecules connected by phosphate groups, with one of the four bases, A, T, C, or G, attached to each deoxyribose. The repeating units of DNA, base-deoxyribose-phosphate (A, T, C, or G) are referred to as *nucleotides*. The exact position of every atom in each of the four nucleotides was known from the research of Lord Alexander Todd and others. For his research on the chemistry of DNA, Lord Todd received the 1957 Nobel Prize in Chemistry.

Another experimental finding crucial to solving the structure of DNA was Chargaff's rules (discussed in Chapter 2): A = T and C = G. Chargaff claimed that he met Watson and suggested that A must be close to T and G close to C in the 3D structure of DNA. Watson did not mention this conversation in his book or in any of his other publications.

Of central importance to Watson and Crick were the X-ray diffraction photographs of DNA produced and interpreted by Rosalind Franklin at King's College, London. Experts in this field claim that Franklin's pictures of DNA were not only the best X-ray photographs of DNA but the most beautiful photographs of any substance ever taken. X-ray crystallography is essentially a form of very high resolution microscopy. In all forms of microscopy, the amount of detail or the resolution is limited by the wavelength of the electromagnetic radiation used. With light microscopy, where the shortest wavelength is about 300 nm (1 nm is a billionth of a meter), one can see bacteria. With electron microscopy, where the wavelength may be 10 nm, one can see viruses and the shapes of large protein and DNA molecules. In order to see DNA in atomic detail, it is necessary to work with electromagnetic radiation with a wavelength of around 0.1 nm, which corresponds to X-rays.

In light microscopy, the subject is irradiated with light and causes the incident radiation to be diffracted in all directions. The diffracted beams are then collected, focused, and magnified by the lenses in the microscope to give an enlarged image of the object. Unfortunately, it is not possible to physically focus an X-ray diffraction pattern, so it has to be done mathematically and this is where the computers come in. The diffraction pattern is recorded using some sort of detector, such as X-ray sensitive film. The X-rays are diffracted by the electrons in the structure and consequently the result of an X-ray experiment is a three-dimensional map showing the distribution of electrons in the DNA structure.

Franklin's pictures showed that DNA was in the form of a cylindrical helix (spiral) with a pitch of 3.4 nm and a diameter of 2 nm. Furthermore, the pictures showed that the helix was composed of more than one chain and the phosphate groups were near the outside of the cylinder. The data were presented by Franklin in a colloquium that took place at King's College in November 1951, 1 year before Watson and Crick published their classic paper in *Nature* on the double helix structure of DNA. Watson was present at the colloquium, but writes in *The Double Helix* that he did not understand Franklin's lecture. He also incorrectly claimed that Franklin was definitely opposed to a helix structure.

Between 1951 and 1953 Rosalind Franklin came very close to solving the DNA structure. She was beaten to publication by Crick and Watson in part because of the friction between Maurice Wilkins and herself. Wilkins was working independently on DNA at King's College. Without getting her knowledge or permission, Wilkins showed Watson and Crick one of Franklin's crystallographic portraits of DNA. When they saw the picture, the solution became apparent to them, and the results went into their *Nature* article almost immediately.

Some years later Wilkins expressed his opinion on the timing of the discovery of the double-helix: "it was all here. They were working at Cambridge along certain lines. And we were working along certain lines. It was a question of time. They could not have gone on to their model, their correct model, without the data developed here. They had that—I blame myself, I was naïve—and they moved ahead. Put it this way, if they (Watson and Crick) were out of the picture entirely, we would still have got it, though it would have taken a bit longer. If we were out of the picture, if they hadn't gotten our stuff, they'd have had to develop it, and that would have taken time—I don't know how long, I think longer still. We were scooped, I don't think fairly."

When Crick was asked in 1970 if he believed that someone in King's college would have solved the problem if he and Watson hadn't, Crick said, "Oh don't be silly. Of course, Rosalind would have solved it. With Rosalind it was only a matter of time."

The story of the Watson and Crick discovery of the double-helix structure of DNA begins in the fall of 1951 when, at the age of 24, Jim Watson arrived at the Cavendish Laboratory in Cambridge, England. He had received his BSc degree in Zoology from the University of Chicago, PhD degree from Indiana University in Molecular Biology and had already carried out 1 year of postdoctoral research in Copenhagen on bacterial viruses. On his first day in Cambridge, Watson met Francis Crick and immediately struck up a friendship with him, based primarily on their joint interest in DNA. Although already 35 years old at the time, Crick had not yet received his PhD degree. He had studied physics at University College, London, obtaining a BSc in 1937. His PhD studies were interrupted by the outbreak of the war in 1939. During the war he worked as a scientist for the British Admiralty, mainly in connection with magnetic and acoustic mines. After the war, Crick became interested in biology and studied organic chemistry and protein crystallography. Scientists at the Cavendish Laboratory quickly realized that Crick had an exceptional mind and frequently sought his advice on difficult problems.

The starting point for Watson and Crick was the recent discovery by Linus Pauling in Pasadena, California of the single-stranded helix structure of proteins. As Watson wrote, "within a few days after my arrival, we knew what to do: imitate Linus Pauling and beat him at his own game." Indeed, Crick and Watson feared that they would be upstaged by Pauling, who proposed his own model of DNA in February 1953, although his three-stranded helical structure quickly proved erroneous.

Pauling, then the world's leading physical chemist, had pioneered the method of combining X-ray crystallography with model building in chemistry. Watson and Crick prepared cardboard and wire models of the four nucleotides, such that the relative size and shape of each of the components fit the known parameters. However, there are so many different ways to arrange the nucleotides into a large molecule that it is necessary to obtain hints on the structure from X-ray photographs. The initial photos they obtained of DNA fibers from Wilkins and Franklin showed an X shape, which is characteristic of a helix. Accordingly, Watson and Crick connected their cardboard nucleotides to form a helix with the bases facing out (so they could interact with proteins). When Franklin was asked to view this model, she told them it could not be correct because the X-ray photographs indicate that the phosphates were located on the outside and the bases in the interior of the helix.

Accepting Franklin's criticism, Watson and Crick began building DNA models with the sugar-phosphate backbone on the outside and the bases in the inside. At this point, they made an insightful discovery: if the cylindrical helix was to maintain a constant diameter, then A must be adjacent to T and G adjacent to C. The reason for this is the geometry of the bases, A and G are purines containing two fused rings, whereas T and C are smaller pyrimidines containing a single ring.

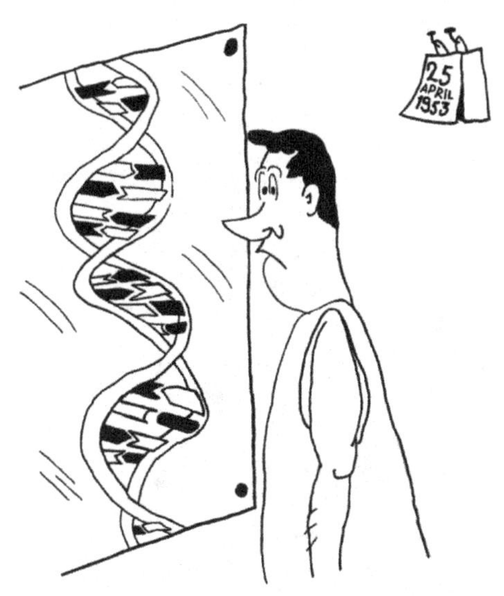

FIGURE 4.1 **Double stranded helix of DNA, showing base pairing between purines and pyrimidines.**

To keep a constant diameter a pyrimidine must border a purine. The most chemically stable arrangement is to place A next to T and C next to G (Figs. 4.1 and 4.2).

Once it was understood that A was next to T and C next to G on the inside of the molecule, it became clear that DNA must be a double helix, with A on one strand adjacent to T on the other strand, and G adjacent to C. The double helix model, with the bases pointing inside and perpendicular to the backbone, fit the X-ray results of Franklin, the known chemistry of the nucleotides, and Chargaff's rules. The discovery of the DNA structure was a landmark achievement. It immediately stimulated studies in many areas of biology, including biochemistry, genetics, evolution, developmental biology, and molecular biology.

Watson and Crick published the structure of the DNA helix in the journal *Nature* in April, 1953. In the same issue, Franklin and Wilkins published their X-ray photographs of DNA. Nine years later, in 1962, Watson, Crick, and Wilkins shared the Nobel Prize in Physiology or Medicine, for solving one of the most important of all biological riddles. Many voices have argued that the Nobel Prize should also have been awarded to Rosalind Franklin, since her experimental data provided the central evidence leading to the solution of the DNA structure. In a recent interview in the magazine Scientific American, Watson himself suggested that it might have been a good idea to give Wilkins and Franklin the Nobel Prize in Chemistry, and Crick and him the Nobel Prize in Physiology or Medicine—in that way all four would have been honored. Rosalind Franklin died in 1958. As a rule only living persons can be nominated for the Nobel Prize, so the 1962 Nobel Prize was out of the question.

One of the exciting aspects of the Watson–Crick model of DNA is that it immediately suggested an hypothesis for the replication of genetic material, so

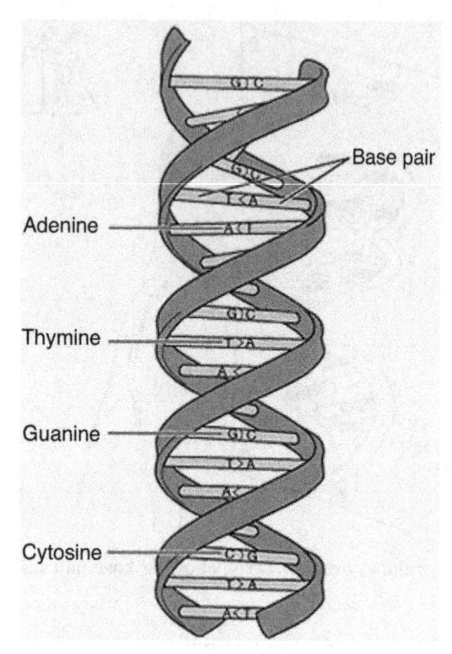

FIGURE 4.2 **Drawing of the double-strand structure of DNA.**

that it can accurately be passed down from generation to generation. In typical British understatement, Watson and Crick wrote in the next to last sentence of their classic 1953 paper, "it has not escaped our notice that the specific pairing we have postulated immediately suggests a possible copying mechanism for the genetic material." What Watson and Crick meant was that the accurate replication of DNA is achieved by the two complimentary stands separating, and each strand serving as a template for the synthesis of a new complimentary strand. In this way, two identical double helix DNA molecules are created which then can be transferred to two newly divided cells. Support for this hypothesis is presented in the next chapter (Fig. 4.3).

- *James Watson* returned to the United States in 1956 to become a Harvard faculty member, and wrote a useful introductory textbook on molecular biology, *Molecular Biology of the Gene.* In 1968, he became the Director of the Cold Spring Harbor Laboratory of Quantitative Biology in New York. In 1990, Watson was appointed as the Head of the Human Genome Project at the National Institutes of Health, a position he held until April 10, 1992. Watson left the Genome Project after conflicts with the new NIH Director, Bernadine Healy. In 2007, he was relieved of his administrative responsibilities at the Cold Spring Harbor Laboratory after publically claiming that black people were less intelligent than white people and the idea that "equal powers of reason" were shared across racial groups was a delusion.

FIGURE 4.3 (A) James Watson (1928–), (B) Francis Crick (1916–2004), (C) Rosalind Franklin (1921–1958), and (D) Photograph of Maurice Wilkins (1916–2004). *(Taken from https://commons.wikimedia.org/wiki/Category:Maurice_Wilkins#/media/File:Maurice_Wilkins_nobel.jpg.)*

- *Francis Crick* completed his PhD in 1954 on a thesis entitled "X-ray diffraction: polypeptides and proteins." He then went to the Brooklyn Polytechnic Institute, where he continued to develop his skills in the analysis of X-ray diffraction photographs of proteins. During the mid-to-late 1950s Crick was intellectually engaged in sorting out the mystery of how proteins are synthesized from the genetic information stored in DNA molecules. Crick became a highly influential theoretical molecular biologist, contributing to the solution of many problems, including the genetic code. In 1976, Crick left Cambridge in order to work at the Salk Institute in California in the field of neuroscience, with special interest in consciousness. In 1995, he wrote an interesting book, *The Astonishing Hypothesis: The Scientific Search for the Soul*. He died in 2004.

- *Rosalind Franklin* left King's College and DNA research shortly after the publications in Nature. She was not happy at the College, partly because her feud with Wilkins and partly because of chauvinistic policies of the College. For example, she was forced to eat in the student's cafeteria because women were not allowed in the all men's Faculty Club. She obtained a position at Birkbeck College in London and between 1953 and 1958 published 17 excellent articles on the structure of viruses. She became ill with ovarian cancer in 1956 and died 2 years later at the age of 37, without an award, but not without recognition—eventually. In recent years Rosalind Franklin has become a kind of feminist icon.
- *Maurice Wilkins* continued to work on the structure of DNA and RNA at King's College, London. He became Assistant Director of the Medical Research Council Unit in 1950 and Deputy Director in 1955. A subdepartment of Biophysics was formed in King's College, and he was made Honorary Lecturer in it. In 1961, a full Department of Biophysics was established. He was elected a Fellow of the Royal Society in 1959, given the Albert Lasker Award (jointly with Watson and Crick) by the American Public Health Association in 1960, and made Companion of the British Empire in 1962. In 1969, Wilkins became the founding President of the British Society for Social Responsibility in Science. In 2000, King's College London opened the Franklin-Wilkins Building in honor of Dr. Franklin's and Professor Wilkins' work at the college. The wording on the base of a DNA sculpture (donated by James Watson) outside Clare College, Cambridge, England is:
  - "These strands unravel during cell reproduction. Genes are encoded in the sequence of bases."
  - "The double helix model was supported by the work of Rosalind Franklin and Maurice Wilkins."

NOTES AND REFERENCES

27 The Modern Library placed *The Double Helix* by James Watson at number 7 on its list of the 20th century's best works of non-fiction.

27 The Nobel Laureate Linus Pauling wrote that Anne Sayre's autobiography of Rosalind Franklin is "A fine book – it is well-written and illuminating in its account of Rosalind Franklin's contributions to the discovery of the double-helix structure of DNA."

31 Watson, J.D., Crick, F.H.C., 1953. A structure for deoxyribose nucleic acid. Nature 4356, 737.

31 The most stable pairs form the maximum number of H bonds (chemical bonds consisting of a hydrogen atom between two electronegative atoms, e.g., oxygen or nitrogen).

32 Watson was opposed to Healy's attempts to acquire patents on gene sequences, and any ownership of the "laws of nature." Two years before stepping down from the Genome Project, he had stated his own opinion on this long and ongoing controversy which he saw as an illogical barrier to research; he said, "The nations of the world must see that the human genome belongs to the world's people, as opposed to its nations."

Chapter 5

Duplicating DNA

The most incomprehensible thing about science is that it is comprehensible.

—Albert Einstein

Stand and unfold yourself.

—William Shakespeare, *Hamlet*

The Watson–Crick Model of DNA revolutionized biology and biochemistry, not so much because it explained a large body of data on the structure of DNA, but because it immediately suggested (to Watson and Crick) how DNA might create an identical copy of itself. As Maurice Wilkins explained, "It is essential for genetic material to be able to make exact copies of itself; otherwise growth would produce disorder, life could not originate, and favorable forms would not be perpetuated by natural selection."

Since the structure of DNA consists of two complimentary strands, either chain thereby carries the information for making the entire molecule. If you know the sequence of bases down one strand of the helix, you always know the sequence of the bases on the other strand. This complementarity is the fundamental reason why a DNA helix can be split down the middle, and then have the other half perfectly recreated. The mechanism that Watson and Crick proposed for DNA replication was elegant in its simplicity: *the two strands separate and each strand serves as template for the synthesis of its partner.*

Consider a segment of DNA consisting of six nucleotides on each strand. When the strands separate, each of the original strands acts as a mold or template for the production of a new chain. A strand having the sequence ACGATT is copied to yield a fresh strand that must have the sequence TGCTAA because of the rigorous base-pairing requirements imposed by the DNA structure. Likewise, the original TGCTAA strand acts as a template for a new strand with the sequence ACGATT. In this way, two identical sets of genetic information can be produced from the original one (Fig. 5.1).

The Watson–Crick hypothesis of DNA replication was put to the test in 1958 by an ingenious experiment designed and performed by two young scientists at the Caltech, Mathew Meselson, and Frank Stahl. Meselson's PhD thesis involved the development of a technique for separating DNA molecules according to their densities (their mass per unit volume). Meselson found that if certain

It's in Your DNA. http://dx.doi.org/10.1016/B978-0-12-812502-1.00005-6

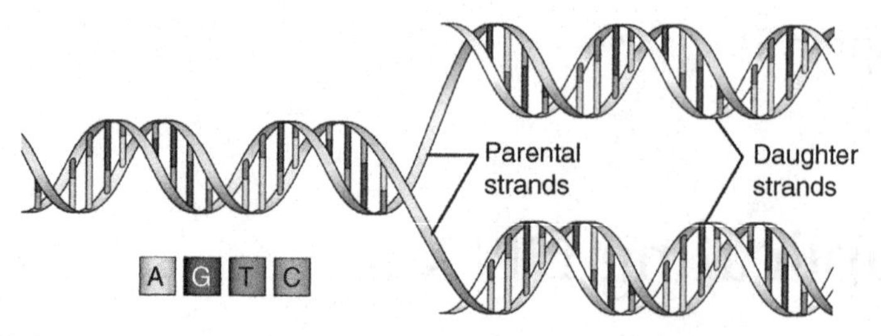

FIGURE 5.1 **The Watson–Crick Model of strand separation and duplication.**

salts, such as cesium chloride (CsCl), are dissolved in water and then centrifuged at very high speeds, the salt molecules are pushed to the bottom of the tube because of centrifugal force. Since CsCl is heavier than water, the higher concentrations of CsCl at the bottom than the top of the tube results in the formation of a gradient of densities, the highest density of CsCl being at the bottom of the tube and the lowest at the top. When DNA is added to the salt solution and then centrifuged, the DNA molecules collect in a thin band at that part of the CsCl gradient that has precisely the same density as DNA. DNA molecules that have even small differences in their densities will concentrate at different positions in the tube and can thereby be separated. The density of a particular DNA molecule depends on the relative amount of bases it contains because the pair G and C is denser than the pair A and T.

When the geneticist Stahl heard about Meselson's technique they met and discussed how the method could be used to test the Watson–Crick DNA replication model. The experiment they designed began by growing the bacterium *Escherichia coli* for a few days in a growth medium that contained the heavy isotope of nitrogen (notated as ^{15}N) in the form of $^{15}NH_3$ (ammonia). As far as the bacteria are concerned, the presence of ^{15}N instead of the usual ^{14}N makes very little difference. The bacteria grow and multiply at the normal doubling time of 30 min. However, when the DNA is extracted from these bacteria, it contains ^{15}N in place of ^{14}N in all of the bases A, T, C, and G. Thus, the DNA is of greater density than the normal ^{14}N-DNA (Fig. 5.2).

After the bacterial DNA had become completely labeled with ^{15}N, the bacteria were transferred to a medium that contained $^{14}NH_3$ and allowed to continue their growth. Henceforth, all DNA would be made with the ^{14}N only. Samples of the bacterial culture were removed at 0, 30, 60, and 90 min after transfer to the ^{14}N medium, and the DNA extracted, mixed with CsCl solution and centrifuged. The results are shown later (Fig. 5.3).

At the beginning of the experiment (time zero), the entire DNA concentrated near the bottom of the tube, corresponding to ^{15}N-DNA. After 30 min (one generation), the entire DNA had a density that was intermediate between ^{15}N-DNA and ^{14}N-DNA, which will be referred to as hybrid DNA. After 60 min (two generations), the DNA was 50% hybrid and 50% ^{14}N-DNA, and after 90 min (three generations), the DNA was 25% hybrid and 75% ^{14}N-DNA. The results are shown in Fig. 5.3.

FIGURE 5.2 Bacteria growing on $^{15}NH_3$ produce heavy (dense) DNA.

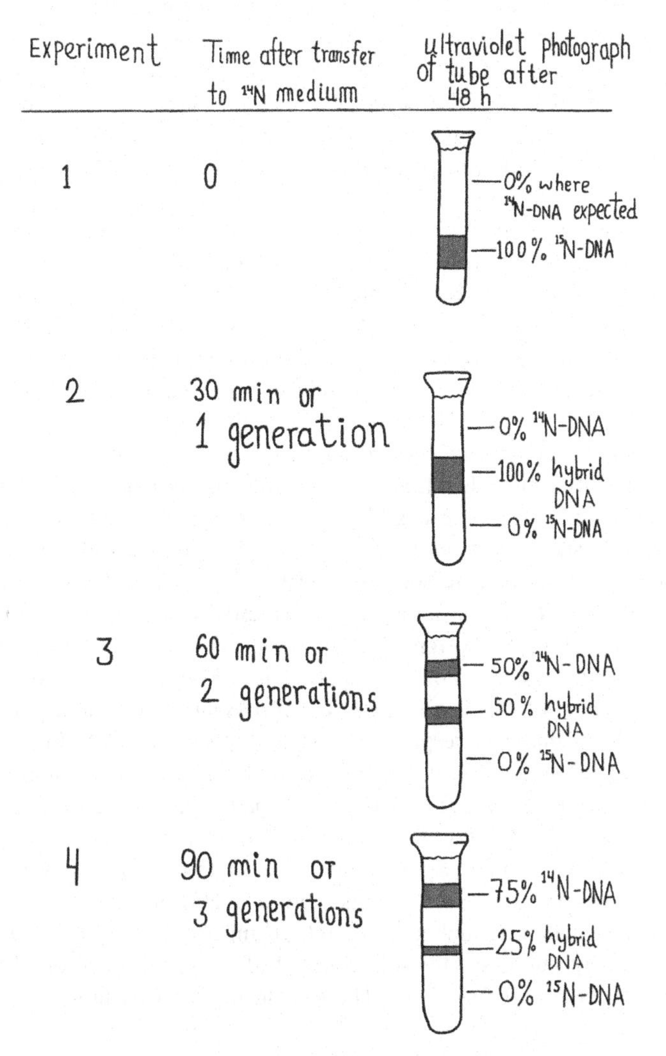

FIGURE 5.3 The results of the Meselson–Stahl experiment. The methods are described in the text.

As discussed briefly in Chapter 2, the nature of science is such that experiments can disprove hypotheses but they cannot prove them. It is always possible that there is an alternative hypothesis that is consistent with the data. Thus, the Meselson–Stahl experiment did not prove the Watson–Crick hypothesis, but it did make the hypothesis more acceptable, and it did eliminate several alternative hypotheses. For example, it had been proposed earlier that the entire double-stranded DNA molecule acts as a template for new DNA. This hypothesis predicts that after one generation in ^{14}N medium, half of the DNA should be the original conserved ^{15}N-DNA and half should be ^{14}N-DNA. Since the entire first-generation DNA had the density of the hybrid, this alternative hypothesis is not consistent with the data and therefore was discarded. To date, the Watson–Crick replication hypothesis is consistent with all the facts and is applicable to the replication of DNA of all living organisms, from the simplest virus to the most complex animals and plants.

The Watson–Crick Model of DNA replication indicates how the sequences of bases in new chains of DNA are specified by the parental DNA: the information comes from the "old DNA." But precisely how are the new chains assembled and where does the energy for executing the process come from?

While work was proceeding on the structure of DNA and the mode by which it replicates in the cell (in vivo), a group at Washington University in St. Louis, under the leadership of Arthur Kornberg, was examining the possibility of replicating DNA in vitro (that is, outside the living cell, in a tube). Kornberg was born in Brooklyn, New York in 1918 and educated in its public schools. He received his undergraduate degree in science from the City College of New York in 1937 and the MD degree from the University of Rochester in 1941. After a year's internship in internal medicine, he served as a commissioned officer in the US Public Health Service. He was first assigned to the Navy as a ship's doctor, and then as a research scientist at the National Institutes of Health (NIH) in Bethesda, Maryland, from 1942 to 1953. He obtained training in the study of enzymes from Professor Severo Ochoa at New York University School of Medicine in 1946. Upon returning to Bethesda, he organized and directed the Enzyme Section. He resigned in 1953 with the rank of Medical Director, to assume the chairmanship of the Department of Microbiology at Washington University School of Medicine in St. Louis, Missouri, where he began experimenting with the enzymes responsible for the synthesis of DNA (Fig. 5.4).

Kornberg reasoned that the minimum requirements for the biosynthesis of DNA molecules were: (1) the component parts, that is, deoxyribose, phosphate, and the four bases, A, T, C, and G; (2) energy to assemble the component parts; (3) DNA to act as a template; and (4) an enzyme catalyst to speed up the process. The component parts were available in several biochemical supply houses that store and sell research chemicals. To facilitate the experiments, Kornberg purchased the four bases, A, T, C, and G, each of which was already connected to deoxyribose. These molecules that contain one base and

FIGURE 5.4 Photograph of Arthur Kornberg (1918–2007). *(From U.S National Library of Medicine, Images from the History of Medicine Collection. Taken from https://commons.wikimedia.org/wiki/File:Arthur_Kornberg.jpg.)*

one deoxyribose are called deoxyribonucleosides. The four deoxyribonucleosides they used in their experiments are abbreviated dA, dT, dC, and dG. As an energy source, he used commercially available adenosine triphosphate (ATP), the source of energy for almost all biological reactions. Items 3 and 4 were supplied by extracts of bacteria, prepared by grinding the bacteria with powdered glass in a mortar with a pestle. This process disrupts the bacteria and releases the DNA and enzymes of the cell. This, then, was how they obtained the four ingredients for their initial experiments (Fig. 5.5).

The next question was how to measure the newly synthesized DNA. Initially they expected to make very little DNA, if any at all; thus they needed an extremely sensitive test. The method they used made use of radioactive tracers. The initial Kornberg experiment consisted of incubating at 37°C a mixture of the four deoxyribonucleosides (one of them, ^{14}C-dT, which he obtained from a colleague in the Pharmacology Department, was radioactive), ATP, and a bacterial extract containing DNA and enzymes. After incubation for about 1 h, cold acid was added to the mixture. They knew from previous studies that free deoxyribonucleosides are soluble in cold acid, whereas DNA precipitates in the acid, so that any radioactivity found in the precipitate indicates that DNA was synthesized during the incubation period. Although only about 0.01% of the radioactivity that was present in the ^{14}C-dT was found in the acid precipitated DNA. Kornberg was confident that the results were significant for two reasons: (1) control experiments in which he left out any one of the ingredients of the mixture resulted in no detectable radioactivity in the acid precipitate, meaning that no new DNA was made, and (2) treatment of the product with the enzyme DNase, which breaks down DNA and had just then become commercially

FIGURE 5.5 **Kornberg and his associates synthesizing DNA from the four nucleoside triphosphates.**

available, rendered all of the radioactivity acid-soluble. This experiment indicated for the first time that *DNA could be synthesized in a test tube*.

For the next 10 years, Kornberg and his associates performed a series of brilliant experiments that demonstrated in detail how DNA was synthesized. First, the four deoxyribonucleosides were converted to deoxyribonucleoside triphosphates, dAppp, dTppp, dCppp, and dGppp, by enzymes and ATP. Once these four compounds are synthesized there is no longer a need for the energy that comes from ATP because that was supplied by the four deoxynucleoside triphosphates. Next, the enzyme that polymerizes the deoxyribonucleoside triphosphates, referred to as DNA polymerase 1, was purified. Finally, using the purified DNA polymerase 1, a mixture of the four deoxribonucleoside triphosphates and template DNA, Kornberg could synthesize in vitro substantial amounts of DNA, the characteristics of which were determined by the template DNA. For example, if he used DNA isolated from a particular virus as the template, the viral DNA was synthesized even though the DNA polymerase came from a bacterium. In essence, DNA polymerase 1 acts like a Xerox machine, accurately duplicating the information stored in the sequence of bases in DNA.

To further demonstrate the accuracy of DNA polymerase 1 in duplicating DNA, Kornberg used the small spherical virus ϕX-174 that infects bacteria (termed therefore—bacteriophage, or phage for short). The DNA of this phage is unusual in several ways. At the time of Kornberg's experiment, ϕ-X174 DNA was the smallest known naturally occurring DNA molecule containing only 5,500 bases. Rather than being double stranded, the DNA isolated from phage ϕ-X174 contains only a single strand of DNA. Most important of all, under appropriate conditions the DNA of ϕ-X174, by itself, can infect bacteria, multiply in the cells and then break open the cells releasing several hundred new complete ϕ-X174 phages. Kornberg demonstrated that DNA polymerase 1 could reproduce ϕ-X174 DNA faithfully, such that the newly synthesized DNA could infect bacteria. Some scientists have claimed that this experiment shows that *life* can be produced in a test tube.

For his pioneer research on DNA synthesis, Arthur Kornberg received the 1959 Nobel Prize in medicine and physiology. Interestingly, his son, Roger David Kornberg, followed in his father's footsteps and received the Nobel Prize in chemistry in 2006 for his studies of the process by which genetic information from DNA is translated to RNA.

Arthur Kornberg is also known for creating the *Ten Commandments of Enzymology*. Unlike the author of the original Ten Commandments, Kornberg was able to modify and amend his commandments as new developments came along.

Thou shalt…

- **I.** Rely on enzymology to resolve and reconstitute biologic events
- **II.** Trust the universality of biochemistry and the power of microbiology
- **III.** Not believe something just because you can explain it
- **IV.** Not waste clean thinking on dirty enzymes
- **V.** Not waste clean enzymes on dirty substrates
- **VI.** Use genetics and genomics
- **VII.** Be aware that cells are molecularly crowded
- **VIII.** Depend on viruses to open windows
- **IX.** Remain mindful of the power of radioactive tracers
- **X.** Employ enzymes as unique reagents

More recent studies on DNA biosynthesis have revealed that the details are considerably more complicated than originally supposed. The enzyme Kornberg studied extensively is not solely responsible for DNA synthesis in living cells. Several different enzymes are necessary for DNA replication in growing cells. The reason for this has to do with two aspects of the structure of DNA that I failed to mention in Chapter 3. The first is that the two strands that make up the double helix run in opposite directions. On one strand, carbon 5 of the deoxyribose sugar is attached to the phosphorus which in turn is attached to carbon 3 of the next sugar, so that the direction is termed 5 to 3. On the other strand, the direction is 3 to 5. However, DNA polymerase only works in one direction, 5 to 3. Since both DNA strands in the chromosome elongate together, it follows that one strand has to be made in the opposite direction in small pieces and then connected.

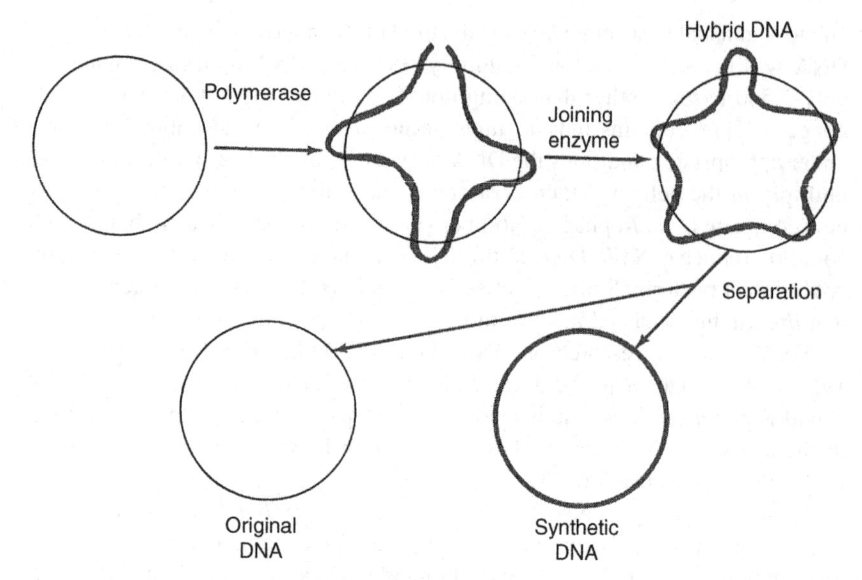

FIGURE 5.6 The role of DNA ligase in DNA replication.

The small fragments of DNA were actually discovered by the Japanese molecular biologist Reiji Okazaki, and accordingly referred to as Okazaki fragments. Okazaki was born in Hiroshima, Japan. He died of leukemia in 1975 at the age of 44; he had been heavily irradiated in Hiroshima when the first atomic bomb was dropped on August 6, 1945.

The enzyme that connects the small pieces of DNA is called DNA ligase. DNA ligase puts together the DNA fragments. The DNA polymerase and DNA ligase work in harmony so that the actual number of small DNA pieces at any time is very low. A diagram showing the role of DNA ligase in DNA replication is presented in Fig. 5.6.

In addition to playing a role in DNA duplication in cells, DNA ligase is one of the most important implements used in genetic engineering. Any construction in which a new piece of DNA (gene) is introduced into an organism requires that the transplanted DNA becomes linked to the existing genetic material. DNA ligase is used both in vivo (in cells) and in vitro (in tubes) to perform this essential reaction. Examples of the use of DNA ligase in genetic engineering will presented in Chapter 10.

The second aspect of DNA structure that demands the action of another enzyme for DNA replication is the coiled nature of the DNA double helix, which is a billion times longer than it is wide. A single human cell contains enough DNA to stretch to about 2 m if it were not tightly coiled. Human DNA is divided into 46 molecules (one long DNA molecule for each chromosome). Since the human body consists of approximately 10^{13} cells, the total length of DNA in a single person is about 2×10^{13} m. Some idea of the extreme length of this DNA can be obtained by comparing it with the distance from the earth to the sun, which is 1.5×10^{11} m. You can see that the DNA in your body could stretch to the sun and back about 50 times. This staggering fact emphasizes the point that the DNA must be tightly packed to fit into cells.

For DNA to be copied by DNA polymerase 1, the two strands of the helix must be separated. The enzyme that performs the job, referred to as DNA helicase, first binds to a specific site on the DNA, the place where DNA synthesis begins. Bacteria have only a single site on their chromosome where DNA synthesis begins, while higher organisms like humans have several on each of their chromosomes. Once DNA helicase binds to the DNA, it separates the two strands—this allows DNA polymerase 1 to attach and begin copying the DNA. As the replication process continues, the helicase moves down the DNA and continues separating the DNA strands.

Kornberg's pioneering research on DNA polymerase has led to today's DNA synthesis techniques and technologies becoming a cornerstone of modern molecular biology and playing a fundamental role in many fields of basic and applied biological research. The ability to synthesize whole genes, novel genetic pathways, and even entire genomes is no longer a dream. Using little more than some commercially available equipment, readily available small pieces of DNA with defined sequences and the four nucleoside triphosphates, and DNA polymerases, a standard molecular biology laboratory can synthesize several thousand base pairs of synthetic DNA in a few days using existing techniques.

Laboratory synthesis of DNA plays an essential role in analyzing forensic biological evidence, determining evolutionary relationships between living and fossil animals and plants, rapid detection of pathogens in medical microbiology, producing various human proteins in microorganisms, including insulin and growth hormones, producing safe and effective bacterial and viral antigens, and probing for specific disease genes, including oncogenes (genes that are connected to cancer development). In Chapter 10 we will discuss genetic engineering and some of these DNA technology applications.

NOTES AND REFERENCES

36 Meselson, M., Stahl, F.W., 1958. The replication of DNA in *Escherichia coli*. Proc. Natl. Acad. Sci. 44, 671–682.

40 Lehman, I.R., Bessman, M.J., Simms, E.S., Kornberg, A.J., 1958. Enzymatic synthesis of deoxyribonucleic acid. I. Preparation of substrates and partial purification of an enzyme from *Escherichia coli*. Biol. Chem. 233, 163–170.

40 Bessman, M.J., Lehman, I.R., Simms, E.S., Kornberg, A.J., 1958. Enzymatic synthesis of deoxyribonucleic acid. II. General properties of the reaction. J. Biol. Chem. 233, 171–177.

41 The smallest virus known today, however, is the single stranded DNA virus *Porcine circovirus* type 1. It has a genome of only 1759 nucleotides.

41 Kornberg, A., 2003. Ten commandments of enzymology, amended. Trends Biochem. Sci. 28, 515–517.

42 Sugimoto, K., Okazaki, T., Imae, Y., Okazaki, R., 1969. Mechanism of DNA chain growth. 3. Equal annealing of T4 nascent short DNA chains with the separated complementary strands of the phage DNA. Proc. Natl. Acad. Sci. USA 63, 1343–1350.
42 Zheng, L., Shen, B., 2011. Okazaki fragment maturation: nucleases take centre stage. J. Mol. Cell Biol. 3, 23–30.
43 Wu, Y., 2012. Unwinding and rewinding: double faces of helicase? J. Nucleic Acids Article ID 140601, 14 p.

Chapter 6

From Genes to Enzymes

It seems reasonable to suppose that the gene's primary and possibly sole function is in directing the final configurations of protein molecules.

—George Beadle

Protein synthesis is a central problem for the whole of biology, and that it is in all probability closely related to gene action.

—Francis Crick

In Chapter 3, evidence was presented that demonstrated that the genetic material is DNA. Then, the double helical structure of DNA was described (Chapter 4), which led to a relatively simple model of DNA duplication (Chapter 5). In this chapter, I begin to address the crucial question: how is the genetic information translated by living organisms into observable physiological characteristics? For example, how does a "gene for blue eyes" actually cause an individual to exhibit that characteristic pigment?

In 1902, the English physician Archibald Edward Garrod published in the journal *Lancet,* the first of a series of articles that dealt with the physiological defect called alkaptonuria. This rare disease, which has been known to the medical community for more than 300 years, is diagnosed by the blackening of the patient's urine on exposure to air. Usually the condition is noticeable in infants from the discoloration of soiled diapers. Alkaptonuria shows a very low prevalence (1:100,000–250,000) in most ethnic groups. One notable exception is in Slovakia, where the incidence of the disease rises to 1:19,000.

Garrod began his study by isolating from the urine of patients with alkaptonuria the chemical that turns black on prolonged contact with air. The agent was purified and identified as homogentisic acid and shown to be present in the urine of patients with the disease but absent from normal individuals. Garrod traced the origin of homogentisic acid to certain foods in the diet, namely the amino acids, tyrosine, and phenylalanine. Amino acids are the building blocks of proteins. Tyrosine and phenylalanine are broken down to produce homogentisic acid in all humans. Normal individuals continue to breakdown homogentisic acid to form carbon dioxide and water as end products. Those afflicted with alkaptonuria, however, cannot breakdown homogentisic acid and therefore

It's in Your DNA. http://dx.doi.org/10.1016/B978-0-12-812502-1.00006-8

excrete it in their urine. The greater the amount of tyrosine and phenylalanine in their diet, the greater the quantity of homogentisic acid excreted.

The fact that alkaptonuria could be detected shortly after birth suggested to Garrod that the disease was inherited. Examination of the marriage records of families having children with alkaptonuria supported this hypothesis. Garrod found that a much larger than expected number were first-cousin marriages. In a lecture Garrod presented to the Royal College of Physicians in 1908, he suggested that the form of heredity of alkaptonuria was best explained by the recently rediscovered Mendelian theories of inheritance of a recessive trait. For the alkaptonuria disease to be expressed, the offspring must acquire the defective gene from both the mother and father. Garrod unveiled the concept and the term "inborn error of metabolism." He wrote: "We may further conceive that the splitting of the benzene ring (in homogentisic acid) in normal metabolism is the work of a special enzyme, that in congenital (from birth) alkaptonuria this enzyme is wanting." In other words, Garrod was stating for the first time a relationship between genes and enzymes.

In spite of Garrod's attempts to popularize his theory of gene action via enzymes, he failed to arouse the interest of biochemists and geneticists sufficiently to follow through on his pioneer studies. It may have been that biochemistry and genetics, both young sciences at the time, did not have the necessary tools to probe more deeply into the problem. The human organism is a difficult subject to study genetically and biochemically—a life cycle that is long, offspring too few, and severe limitations on subjecting humans to chemical analysis in addition, of course, to serious ethical issues. It may also have been that Garrod, like Mendel, was so far ahead of his time that his contemporaries were not ready to consider seriously his far-reaching gene-enzyme concept.

A hint of Garrod's philosphy of science can be seen in an excerpt of the address he made to the Abernethian Society in 1912:

> *Scientific method is not the same as the scientific spirit. The scientific spirit does not rest content with applying that which is already known, but is a restless spirit, ever pressing forward towards the regions of the unknown, ... it acts as a check, as well as a stimulus, sifting the value of the evidence, and rejecting that which is worthless, and restraining too eager flights of the imagination and too hasty conclusions.*

The concept that genes are responsible for the synthesis of specific enzymes is generally attributed to the collaborative research of two Stanford scientists, George Beadle and Edward Tatum, who shared the 1958 Nobel Prize in Physiology or Medicine for these studies (Fig. 6.1).

Beadle grew up on a farm in Nebraska. Thus, it is not surprising that his first research at Cornell University was in the area of corn genetics. In 1931, Beadle was awarded a Fellowship at the California Institute of Technology (CalTech) in Pasadena, where he remained until 1936. During this period, he continued his work on Indian corn and began research on the genetics of the

FIGURE 6.1 (A) George Wells Beadle (1903–89) and (B) Edward Lawrie Tatum (1909–75).

fruit fly, *Drosophila melanogaster*. Beadle was appointed Professor of Genetics at Stanford University in 1937 and there he remained for 9 years, working for most of this period in collaboration with Tatum.

Tatum was born in Colorado and educated at the University of Chicago in chemistry and biochemistry. His PhD degree thesis was on studies of the nutrition and metabolism of bacteria. This research no doubt laid the foundations of his later work with Beadle at Stanford University. During their fruitful collaboration, Tatum took charge of the biochemical aspects of their joint work, whereas Beadle focused on the genetics. Beadle and Tatum's work welded genetics and biochemistry into the new science of biochemical genetics, which itself gave rise to molecular biology and biotechnology. But that is getting ahead of the story.

As genetics developed during the first half of the 20th century, certain organisms came to the forefronts that were well suited for careful genetic and biochemical analysis. One of these was the common red bread mold, *Neurospora crassa*. This fungus possesses many advantages for a study on the mode of gene action. First, it can be maintained easily in the laboratory. Second, starting with a single cell, a population of millions of identical cells can be obtained in a few days. Furthermore, genetic analysis is simplified by the fact that during the greatest part of its life cycle, the bread mold has only a single set of genes (haploid) instead of the two sets (diploid) found in most higher organisms, such as the peas that Mendel had studied (Chapter 3). Thus, complication of dominance and recessiveness is avoided, since gene action cannot be hidden by its dominant counterpart. Equally important, *Neurospora* can be grown in a simple medium consisting of water, various inorganic salts, sugar, and one vitamin of the B group, biotin. From these few substances, the mold can construct all the components of the cell (Fig. 6.2).

FIGURE 6.2 Codiscoverers of the one gene–one enzyme hypothesis.

The ease with which both genetic and biochemical studies could be conducted on *Neurospora* induced Beadle and Tatum to choose that organism for a careful collaborative investigation of the gene–enzyme relationship. As Beadle recollects:

> *In 1940, we decided to switch from* Drosophila *to* Neurospora. *It came about in the following way: Tatum was giving a course in biochemical genetics, and I attended the lectures. In listening to one of these- or perhaps not listening like I should have been- it suddenly occurred to me that it ought to be possible to reverse the procedure we had been following and instead of trying to work out the chemistry of known genetic differences, we should be able to select mutants in which known chemical reactions were blocked.* Neurospora *was an obvious organism on which to try this approach, for its life cycle and genetics had been worked out by Dodge and Lindegren, and it could probable be grown in a culture medium of known composition. The idea was to select mutants unable to synthesize known metabolites, such as vitamins and amino acids which could be supplied in the medium. In this way a mutant unable to make a given vitamin could be grown in the presence of the vitamin and classified on the basis of its differential growth in media lacking or containing it.*

The isolation and characterization of mutants involved the following steps, which are still used today:

1. The mold culture was irradiated with X-rays or ultraviolet light. This increases the frequency of mutation several 1000-fold.
2. A single irradiated fungal cell was placed in a complete medium containing most of the components that are known to be in these cells, such as amino acids, vitamins, and the building blocks of nucleic acids. The cell should

grow in such a medium even if it is a mutant that cannot synthesize one of these components.

3. A sample of the culture was then transferred to a minimal medium that contained only sugar, biotin, salts, and water. As the original *Neurospora* is known to grow on minimal medium, *failure of the mold to grow in this medium is taken as evidence that a mutation has occurred*. Since the mutant *Neurospora* grows on the complete medium but not on the minimal medium, there must be one or more components in the complete medium that the experimentally-produced mutant can no longer synthesize. For example, if a mutation occurred in a gene that is responsible for synthesizing one of the enzymes necessary for the making the amino acid arginine, the cell will not grow unless arginine is added to the medium.
4. The final step was classifying the mutant by identifying the specific component in the complete medium that was necessary for its growth. This was done by testing the mutant for growth in the minimal medium supplemented, in separate cultures, with each of the components in the complete medium. In the case of the example mentioned earlier (arginine mutant), the mutant failed to grow on the minimal medium supplemented with a mixture of all the vitamins, or a mixture of all the nitrogen bases in DNA, T, G, C, and A, but did grow when supplemented with a mixture of the 20 naturally-occurring amino acids. Thus, Beadle and Tatum concluded that the mutation was in the synthesis of one of the amino acids. They then tested each of the 20 amino acids separately. Only the amino acid arginine added to the minimal medium allowed the fungus to grow. The mutant was thus classified as an arginine-requiring mutant.

Using this systematic technique, Beadle and Tatum isolated a large number of different mutant strains of *Neurospora*, each of which required some nutrient to grow. Each of these mutants was characterized by the loss of the capacity to synthesize a specific nutrient. Since all chemical reactions within cells are mediated by enzymes, they concluded that the mutational change resulted in the loss of some enzyme necessary for the biosynthesis of the nutrient in question. Genes must somehow give rise to specific enzymes. This was, of course, what Garrod had suggested (with less evidence) some 40 years earlier.

The Beadle/Tatum experiments, indicating that each gene was responsible for the production of a specific enzyme, came to be called the *one gene-one enzyme hypothesis*. The hypothesis was later modified to "one gene-one protein" because genes are also responsible for the synthesis of a large number of proteins that are not enzymes, such as antibodies that defend against foreign invaders, contractile proteins that are involved in muscle contraction and movement, hormones, such as insulin, and collagens that provide support of tendons and ligaments.

The one gene-one protein concept also had to be modified because many proteins are composed of more than one protein chain, and different genes are responsible for the production of the different protein chains. For example, the blood protein hemoglobin is composed of four chains, two alpha chains and two beta chains. Different genes provide the information for the alpha and beta chains. Thus, the one gene-one enzyme hypothesis then became the one gene-one protein chain hypothesis, or one gene-one peptide hypothesis.

The 19th century biologist Thomas Huxley wrote:

The great tragedy of science is the slaying of a beautiful hypothesis by an ugly fact.

As often occurs in biology, a beautiful hypothesis turns out not to be universally true. The system often becomes more complex when moving from microorganisms, such as bacteria and fungi, to higher organisms, such as animals and plants. We now know that in higher organisms, including humans, one gene can code for two or more different proteins (even up to 1000). The human genome, for example, contains about 21,000 protein-encoding genes, but the total number of proteins in human cells is estimated to be between 250,000 and one million. By a process called splicing parts of the gene information can be eliminated and the remainder spliced or joined together. Consequently, this gives rise to proteins with different amino acid sequences and, often, enzymes with different biological functions. Notably, alternative splicing allows the human genome to direct the synthesis of many more proteins than would be expected from its 20,000 protein-coding genes. Which particular enzymes are produced by splicing is often affected by the environment.

Abnormal variations in splicing are also implicated in disease; a large proportion of human genetic disorders result from splicing variants. One example is Huntington disease, also known as Huntington's chorea, a devastating, late-onset, inherited neurodegenerative disorder that manifests with personality changes, movement disorders, and cognitive decline. It is caused by aberrant splicing of a specific messenger RNA. Abnormal splicing variants are also thought to contribute to the development of cancer.

In the next chapter I will discuss experiments that led to our understanding of how genetic information stored in DNA gives rise to specific proteins, that is, the genetic code.

NOTES AND REFERENCES

45 Bearn, A.G., Miller, E.D., 1979. Archibald Garrod and the development of the concept of inborn errors of metabolism. Bull. Hist. Med. 53, 315–328.

46 Garrod, A.E., 1912. The scientific spirit in medicine: inaugural sessional address to the Abernethian Society. St. Bartholomew's Hospital J. 20, 19.

46 Beadle, G.W., Tatum, E.L., 1941. Genetic control of biochemical reactions in *Neurospora*. Proc. Natl. Acad. Sci. USA 27, 499–506.

48 Beadle, G.W., 1966. Biochemical genetics: some recollections. In: Cairns, J., Stent, G.S., Watson, J.D. (Eds.). Phage and the Origins of Molecular Biology, Cold Spring Harbor Symposia, Cold Spring Harbor Laboratory of Quantitative Biology, New York, NY, pp. 23–32.

49 Hickman, M., Cairns, J., 2003. The centenary of the one-gene one-enzyme hypothesis. Genetics 163, 839–841.

49 Davis, R.H., 2007. Beadle's progeny: innocence rewarded, innocence lost. J. Biosci. 32, 197–205.
50 Pan, Q., Shai, O., Lee, L.J., Frey, B.J., Blencowe, B.J., 2008. Deep surveying of alternative splicing complexity in the human transcriptome by high-throughput sequencing. Nature Genet. 40, 1413–1415.
50 Sathasivam, K., 2013. Aberrant splicing of HTT generates the pathogenic exon 1 protein in Huntington disease. Proc. Natl. Acad. Sci. USA 110, 2366–2370.
50 Fackenthal, J.D., Godley, L.A., 2008. Aberrant RNA splicing and its functional consequences in cancer cells. Dis. Model. Mech. 1, 37–42.

Chapter 7

Cracking the Genetic Code

DNA is like a computer program but far, far more advanced than any software ever created.

—Bill Gates

In describing genetic mechanisms, there is a choice between being inexact and incomprehensible. In making this presentation, I shall try to be as inexact as conscience allows.

—François Jacob, French microbial geneticist and 1965 Nobel Prize recipient

We are now ready to discuss the genetic code, that is, how the sequence of nucleotides in DNA determines the sequence of amino acids in proteins. Unlike DNA replication, where discovery of the double-helix structure led to an immediate hypothesis of how the complimentary strands are duplicated, understanding how DNA controls protein synthesis was more challenging for several reasons. To begin with, proteins are composed of 20 different amino acids, whereas DNA contains only four different nucleotides (each containing one of the four bases: A, T, C, and G). Thus, a one-to-one relationship between the bases in DNA and the amino acids in proteins is theoretically impossible. Second, there is nothing in the chemical structure of amino acids that might suggest how they could interact with the nucleotides in DNA. Third, it was known that most proteins are synthesized in the cytoplasm, whereas DNA was located on chromosomes in the nucleus. It is therefore obvious that an intermediate molecule must carry the genetic information from the nucleus to the cytoplasm. During the period from 1952 to 1968, genetic and biochemical experiments resolved these problems, uncovered the genetic code, and gave rise to the central dogma of molecular biology:

DNA→ RNA→ protein

The theoretical physicist, George Gamow, most noted as the author of popular books on science, including the Mr. Tompkins ... series of books, was one of the first to attempt to solve the problem of how the order of four different kinds of bases in DNA chains could control the synthesis of proteins from amino acids. From a purely theoretical argument, Gamow concluded that it requires more than two bases to code for a single amino acid. Suppose only one base was to be used as a code for an amino acid, then the four bases could specify only four

It's in Your DNA. http://dx.doi.org/10.1016/B978-0-12-812502-1.00007-X

amino acids. Since there are 20 different amino acids found in proteins, the single base code is excluded.

Using 2 bases to code for 1 amino acid yields 4 × 4, or 16 possible combinations:

AC	CC	GC	UC
AA	CA	GA	UA
AG	CG	GG	UG
AU	CU	GU	UU

Since there are not enough doublets to account for the 20 amino acids, Gamow concluded that more than 2 bases code for 1 amino acid. If 3 bases are used in each code word, there are 4 × 4 × 4, or 64, possible triplets, which is more than adequate.

The first experimental support for the concept of a triplet code came from the elegant genetic experiments of Frances Crick and coworkers performed in 1960 at the Cavendish Laboratory in Cambridge, England. The hypothesis that they tested was that proteins are read from the beginning of a gene, such that the first amino acid of the protein is determined by the sequence of the first 3 bases in gene, the second amino acid by bases 4–6, the third by bases 7–9, and so on, until the protein is completed with generally more than 100 amino acids in a particular sequence. The precise order of the amino acids is required for the protein to be functional.

The experimental technique that was used to test the triplet hypothesis relied on the fact that the chemical mutagenic agent proflavine causes the deletion of a single base in DNA. They reasoned that the removal of a single base at the beginning of a gene would throw the entire amino acid sequence out of order because the first amino acid would be determined by bases 2–4, the second by 5–7, etc., (instead of 1–3, 4–6, etc.). A second deletion mutation at the start of the gene would also yield an incorrect amino acid sequence. However, a third deletion would realign the proper coding sequence and give rise to a functional protein. When they performed the experiment on a viral gene, the results were as predicted: removal of one, two, or four bases gave an inactive gene and a virus that could not multiply, whereas after deleting three bases, the gene was functional and the virus was viable. From this set of experiments and others it was concluded that genes are read in a linear fashion from a fixed point in one direction, three letters at a time, with no overlaps (meaning that adjacent triplets do not share a base). I should point out that this experiment was performed before the powerful techniques of DNA and protein sequencing were available.

The major question remained: which specific triplet base sequences code for which amino acids? In 1961, at the Fifth International Congress of Biochemistry in Moscow, a young biochemist from the National Institutes of Health (NIH), Marshall Nirenberg, reported his breakthrough results on protein synthesis that amazed the scientific world and led directly to deciphering the genetic code. Before describing this experiment, it is necessary to discuss an additional nucleic acid, termed ribonucleic acid, or RNA for short.

Following the discovery of DNA in the nucleus of white blood cells by Miescher in 1869, a number of other scientists began to isolate nucleic acid from different tissues and different organisms. By 1920, a definite picture had emerged of two types of nucleic acids. The type that Miescher had isolated from white blood cells and salmon sperm and Levene had isolated from calf thymus contained the sugar deoxyribose and was named deoxyribonucleic acid or DNA for short. The other type, which was originally isolated from yeast contained the sugar ribose and was therefore named ribonucleic acid or RNA for short.

The two types of nucleic acids are chemically similar in that they are large molecules composed of backbone chains of sugar-P-sugar-P-sugar-P-sugar-P-etc., with one of the four nitrogen bases attached to each sugar, ribose, or deoxyribose. In the case of RNA, three of the four bases are the same as in DNA, adenine (A), guanine (G), and cytosine (C). The other base is uracil (U) instead of the DNA-specific thymine (T). Another important chemical difference is that RNA is usually single-stranded, whereas DNA is double-stranded.

Although DNA and RNA have been found in all microorganisms and in all the tissues of animals and plants, they are located in different parts of the cell. DNA is found almost exclusively in the nucleus, whereas RNA is present primarily in the cytoplasm. As we shall see, this reflects the different tasks they perform.

Three major types of RNA have been found: ribosomal RNA (rRNA), messenger RNA (mRNA), and transfer RNA (tRNA). All three types of RNA are found in the cytoplasm of cells and are required for protein synthesis. Ribosomal RNA is associated with proteins to form structures called ribosomes. Ribosomes are the sites at which protein synthesis takes place. The detailed structure of ribosomes has only recently been elucidated by three scientists who shared the 2009 Nobel Prize in Chemistry "for studies of the structure and function of the ribosome," Thomas Steitz, an American working at Yale University; Venkatraman Ramakrishnan, an Indian born, US citizen, working at Cambridge University; and Ada Yonath, an Israeli woman, working at the Weizmann Institute. All three laureates generated 3D models that show, among other things, how different antibiotics bind to bacterial ribosomes and block protein synthesis but do not affect human ribosomes. These models are now being used by scientists in order to develop much-needed new antibiotics effective against human pathogenic bacteria, which are resistant to existing antibiotics.

Messenger RNA comprises a large group of molecules that are synthesized in the nucleus and then moves to the ribosomes in the cytoplasm where they direct protein synthesis. Each messenger RNA has a base sequence complimentary to one strand of a DNA gene. The evidence that RNA is synthesized in the nucleus and subsequently transferred to the cytoplasm came from a simple nucleus transplantation experiment performed by two zoologists, Lester Goldstein and Walter Plaut, at the University of California at Berkeley. They began by labeling RNA with radioactivity using established techniques. Next, the duo employed micromanipulation to carefully remove an individual ameba's nucleus and transfer it to an unlabeled ameba in which the nucleus had also been removed. They reasoned that if RNA originated in the nucleus, radioactivity would, at the end of the experiment, be detected in both nucleus and cytoplasm. In contrast, if RNA did not traffic from nucleus to cytoplasm, no radioactivity would be detected in the cytoplasm at the conclusion of the experiment. Goldstein and Plaut followed the voyage of the radioactive RNA by taking samples at different times following the nuclear transfer, fixing the samples, and exposing the fixed samples to photographic film. The radioactivity exposes the film and shows precisely where the radioactivity is located in the cell. At early time points, essentially all the radioactivity was detected within the nucleus, whereas at later time points, radioactivity was detected in the cytoplasm as well. The radioactive RNA had therefore moved from the nucleus to the cytoplasm, indicating a nuclear origin for RNA.

Transfer RNAs are relatively small molecules, containing 70–90 nucleotides that play a special role in protein synthesis. By comparison, messenger and ribosomal RNAs contain more than 1000 nucleotides. A specific amino acid is enzymatically attached at one end to a transfer RNA in the cytoplasm. Each type of transfer RNA molecule can be attached to only one type of amino acid, so each cell needs many types of transfer RNA. The other end of the transfer RNA molecule contains a three-nucleotide sequence called the anticodon. Thus, each transfer RNA molecule has a specific amino acid attached at one end and the triplet anticodon signaling that specific amino acid on its other end. The transfer RNA containing the amino acid is brought to the ribosome where the transfer RNA-amino acids line up in the order specified by the interaction of the anticodons with the codons on the messenger RNA. The ribosome then catalyzes the connections of adjacent amino acids to form the protein chain. The transfer RNAs are then freed to reattach to new amino acids.

The following points should be emphasized: (1) all three species of RNA are made from the DNA template in the nucleus and then move to the cytoplasm; (2) the ribosome is formed from ribosomal RNA and protein, and becomes programmed for protein synthesis only when a messenger RNA is attached to it; (3) the amino acids enter the programmed ribosome after they have been attached to a specific transfer RNA. The transfer RNA acts as a two-handed molecule, one hand holding onto the specific amino acid, the other binding to a specific triplet sequence of bases on the messenger RNA. In this manner the amino acids are placed in the order coded for by the messenger RNA, which in turn is coded for by the DNA. The scheme may seem complicated, but it certainly is ingenious.

With this background on the three types of RNA, we can return to Marshall Nirenberg's breakthrough experiment that broke the genetic code. Nirenberg was studying protein synthesis in vitro (in a test tube outside of a living cell). He found that he could make protein by mixing (1) the 20 natural amino acids found in proteins, (2) ribosomes, (3) an extract of cells (containing DNA, messenger RNA and transfer RNA), and (4) ATP for energy. The measurement of protein synthesis was based on the fact that amino acids are soluble in cold acid, whereas proteins are insoluble. One of the amino acids he used in his experiments was labeled with radioactivity. After incubation of the mixture for 30 min at 37°C, it was chilled, and cold acid was added to it. The reaction mixture was then centrifuged to separate the soluble and insoluble fractions. Nirenberg found radioactivity in the insoluble fraction, indicating that protein had been synthesized.

Next, Nirenberg carried out the same experiment with one important exception: the enzyme deoxyribonuclease (DNase) was added to destroy the DNA that was present in the cell extract. Protein was still synthesized, indicating that DNA is not necessary for protein synthesis in vitro. However, he noticed that in the presence of DNase, protein synthesis stopped after about 20 min.

Why did the reaction stop? To an empiricist, the question is turned into an experiment designed to test if the reaction stopped because one of components in the mixture became depleted. The following simple experiment was performed by Nirenberg: after 20 min with DNase, add to the incubation mixture in separate test tubes, amino acids, ribosomes, messenger RNA, transfer RNA, and ATP. The results of such an experiment were clear: only the messenger RNA stimulated protein synthesis. He reasoned that during the initial 20 min, the messenger was broken down. Since no template DNA was present (DNase destroyed it), the messenger RNA could not be replenished and protein synthesis ceased.

These experiments paved the way for a comparative study of the effect of various messenger RNAs on protein synthesis. *Once the incubation mixture became depleted of the messenger RNA originally present in the cell extract, the type of protein synthesized depended entirely on the type of added messenger RNA*. Nirenberg and his associates found, for example, that messenger RNA from virus-infected cells produced virus proteins. Then, either by design or by chance (only Nirenberg knows for sure), they decided to test an artificial RNA-like molecule that is called "poly-U." Poly-U is a synthetic polymer that has a structure identical to RNA, except that it contains only one type of base, uracil.

Their experiment revealed that poly-U was a surprisingly efficient messenger for protein synthesis. But if the RNA message was UUUUUUU …, what was the protein product? It turned out that it was a polymer containing only one type of amino acid, phenylalanine. Thus, Nirenberg concluded, "Addition of poly-U resulted in the incorporation of phenylalanine alone into the protein poly-phenylalanine. Poly-U appears to function as a synthetic template, or messenger RNA, in this system. One or more uridylic acid residues appear to be the

TABLE 7.1 The Genetic Code

Amino acid	Codon(s)
Phenylalanine	UUU & UUC
Serine	UCU, UCC, UCA, UCG, AGU, & AGC
Tyrosine	UAU & UAC
Cysteine	UGU & UGC
Leucine	UUA, UUG, CUU, CUC, CUA, & CUG
Valine	GUU, GUC, GUA, & GUG
Proline	CCU, CCC, CCA, & CCG
Histidine	CAU & CAC
Arginine	CGU, CGC, CAG, CGG, AGA, & AGG
Isoleucine	AUU, AUC, & AUA
Threonine	ACU, ACC, ACA, & ACG
Alanine	GCU, GCC, GCA, & GCG
Glutamine	CAA & CAG
Asparagine	AAU & AAC
Lysine	AAA & AAG
Aspartic acid	GAU &GAC
Glutamic acid	GAA & GAG
Tryptophan	UGG
Glycine	GGU, GGC, GGA, & GGG
Methionine (start)	AUG
None (stop)	UGA, UAA, & UAG

code for phenylalanine." Assuming a triplet code, the first word (codon) in the genetic dictionary then becomes UUU = phenylalanine.

The news that Nirenberg's poly-U experiment had determined the first "word" of the genetic code became an international media event. The New York Times reported on Nirenberg's discovery by explaining that "the science of biology has reached a new frontier, leading to a revolution far greater in its potential significance than the atomic or hydrogen bomb." The American journalist and science writer John Pfeiffer asserted that the biggest news story of 1961 "was not the orbiting of the Russian astronauts. It was the cracking of a biological code by Marshall Nirenberg, which governs all the processes of life. This is just as big a breakthrough in biology as Newton's discovery of gravitation in the 17th century was in physics."

With the path clearly prepared by Nirenberg's initial discovery, progress was rapid. Polyadenine RNA sequence (AAAAA ...) coded for the polypeptide poly-lysine and poly-cytosine RNA sequence (CCCCC ...) coded for the

FIGURE 7.1 **Universality of the genetic code demonstrates that modern life traces back to a single ancestor.**

polypeptide polyproline. Therefore, the codon AAA specified the amino acid lysine, and the codon CCC specified the amino acid proline. Artificial RNA polymers were then chemically synthesized containing every possible combination of the four bases. They were then tested for their ability to stimulate protein synthesis in vitro. By analyzing the relative amount of each amino acid incorporated into protein under the guidance of these synthetic RNAs, it was possible to assign codons to every natural amino acid. Especially useful were the RNA molecules synthesized by the chemist Har Gobind Khorana and his associates at the University of Wisconsin (Table 7.1).

By the end of 1968, the entire genetic code was deciphered by matching amino acids to synthetic triplet nucleotides. From an analysis of the data that accumulated, several generalizations emerged (Table 7.1).

1. *The code is degenerate.* Degeneracy means that several different triplets can specify the same amino acid. For example, both UUU and UUC code for phenylalanine.
2. *There are codons for start and stop.* The messenger RNA is not simply read from one end to the other. Instead there are specific codons that determine where reading begins and ends. The codon AUG for the amino acid methionine also signals start. Three codons, UGA, UAA, and UAG, signal stop.
3. *The code is largely, if not entirely, universal.* It is known as "universal" because the same genetic code has been found in all known organisms from viruses and bacteria to plants and animals, including humans. However, like

> many rules in biology there are exceptions, and such is the case with the Genetic Code; small variations in the code exist in certain microorganisms. The near-universality of the genetic code gives us confidence that modern life traces back to a single ancestor. If there was more than one origin of life, and its descendants independently developed the DNA→protein system, it would be very unlikely that all modern species would utilize the same code (Fig. 7.1).

Marshall Nirenberg shared the 1968 Nobel Prize in Physiology or Medicine with Har Gobind Khorana and Robert Holley. Khorana was a chemist who synthesized many of the artificial RNA polymers used to crack the genetic code; Holley was the first to sequence a transfer RNA and determine its structure.

Unraveling the genetic code is equivalent to learning the alphabet. It is still necessary to learn to read—and to comprehend what we read. Buried in the sequence of bases in DNA are the genetic histories of species, for only through changes in the order of bases in DNA are mutation and subsequent evolution possible. To unravel the deep secrets hidden in DNA new biochemical methods were needed. Two of the key methods, DNA amplification and sequencing, are discussed in the next chapter.

NOTES AND REFERENCES

53 Watson, J.D., 2002. Genes, Girls, and Gamow: After the Double Helix. Random House, New York, NY.

54 Crick, F.H., Barnett, L., Brenner, S., Watts-Tobin, R.J., 1961. General nature of the genetic code for proteins. Nature 192, 1227–1232.

56 Goldstein, L., Plaut, W., 1955. Direct evidence for nuclear synthesis of cytoplasmic ribose nucleic acid. Proc. Natl. Acad. Sci. USA 41, 874–880.

In August 1961, Nirenberg and Matthaei published their now-classic essay, "The Dependence of Cell-Free Protein Synthesis in *E. coli* upon Naturally Occurring or Synthetic Polyribonucleotides," in the Proceedings of the National Academy of Sciences. In that same month, Nirenberg presented a version of his findings with the poly-U experiments to a small group of about thirty scientists at the International Congress of Biochemistry in Moscow. Francis Crick, who was in attendance at the Moscow meeting, had heard that Nirenberg had found a clue that might unravel one of the central mysteries of molecular genetics. Crick arranged to have the young scientist deliver his paper again, this time to the assembled body of about a thousand people. By the end of the Moscow conference, the discovery made the obscure and mild-mannered NIH scientist a veritable celebrity.

59 Nirenberg, M., Leder, P., Bernfield, M., Brimacombe, R., Trupin, J., Rottman, F., O'Neal, C., 1965. RNA codewords and protein synthesis, VII. On the general nature of the RNA code. Proc. Natl. Acad. Sci. USA 53, 1161–1168.

60 Jukes, T.H., Osawa, S., 1990. The genetic code in mitochondria and chloroplasts. Experientia 46, 1117–1126.

Chapter 8

DNA Sequencing and PCR

As soon as the right method was found, discoveries came as easily as ripe apples from a tree.

—Robert Koch

It is like a voyage of discovery into unknown lands, seeking not for new territory but for new knowledge. It should appeal to those with a good sense of adventure.

—Frederick Sanger, two time Nobel Prize Laureate for inventing the techniques for sequencing amino acids in proteins and bases in DNA

Progress in science depends on new ideas, new discoveries, and new techniques, and often the three are interdependent. Basic research leads to new discoveries, which inspire new ideas and new techniques. Alternatively, new techniques have led to new discoveries and ideas. Nowhere was this more apparent than in the two techniques, DNA sequencing and *P*olymerase *C*hain *R*eaction (PCR) that were born out of creative ideas and rapidly gave rise to an explosion of fundamental and applied discoveries in DNA bioscience.

The earlier technique to be developed was the DNA sequencing method which was essential for reading the genetic information stored in DNA. This method enables determining the base sequence in a DNA molecule. In 1977, the British biochemist Frederick Sanger and his team at Cambridge developed the first practical DNA sequencing method.

Sanger was born in South West England and studied in the Bryanston School near the town of Blandford Forum in Dorset. His excellent grades won him a scholarship to study science at St John's College, Cambridge. Sanger graduated in 1939 and then did an advanced course in biochemistry. As a Quaker, he was a conscientious objector during the Second World War. He remained at Cambridge and completed his PhD degree in 1943. He then received a small grant from the Medical Research Council to work on protein structure. He chose to study the small protein hormone insulin because of its medical implications and because there was a plentiful supply of the bovine variety. Insulin is the central hormone in glucose metabolism. He spent 10 years carefully identifying small fragments of the insulin molecule and working out their composition to arrive at its structure. By 1953 he had the exact sequence of amino acids for insulin. The significance of Sanger's protein research was realized immediately

It's in Your DNA. http://dx.doi.org/10.1016/B978-0-12-812502-1.00008-1

and he was awarded the Nobel Prize in chemistry in 1958, an unusually short time after publishing his results.

In the early 1960s, Sanger became a senior member at the Medical Research Council's new laboratory for molecular biology in Cambridge. Once Crick and Watson had produced an explanation for how the genetic code was inherited through DNA, it was inevitable that Sanger should apply his flair in sequencing polymers to deciphering the detailed construction of individual genes. Once again, Sanger and his team combined the old with the new to develop an original DNA sequencing method, now known as the "Sanger Method." In 1977, Sanger published the sequence of a virus genome of over 5000 base pairs. Modifications of the Sanger Method are now commonplace in molecular biology.

The Sanger Method is based on the DNA polymerase studies of Arthur Kornberg (see Chapter 5). Four reaction tubes are used. Each reaction tube contains (1) DNA polymerase, (2) the four deoxyribonucleoside triphosphates (dAppp, dGppp, dTppp, and dCppp), (3) a single strand of the DNA (the template) to be sequenced, and (4) a small single-stranded DNA primer that binds to the beginning of the much larger DNA template. A primer is a short strand of DNA that serves as a starting point for DNA synthesis. It is required for DNA replication because DNA polymerases, can only add new nucleotides to an existing strand of DNA. These primers are usually short, chemically synthesized pieces of DNA with a length of about 20 bases. They are hybridized to the target DNA, which is then copied by the polymerase to produce double-stranded DNA molecules. Up to this point, the reaction components are exactly as Kornberg described for in vitro DNA synthesis.

Now for the ingenious modification to the Kornberg procedure that Sanger introduced. He added to each tube a small amount of a modified deoxyribonucleoside triphosphate, referred to as a dideoxyribonucleoside triphosphate. This compound can occasionally enter the newly synthesized DNA in place of the normal nucleotide. However, because of its chemical structure it does not allow DNA synthesis to continue and the reaction terminates at that point. Each of the four reaction tubes contained a different dideoxy compound, ddAppp, ddGppp, ddTppp, or ddCppp. The four tubes are labeled "A," "G," "T," and "C." All four tubes contained DNA polymerase, the DNA template molecule whose sequence is to be analyzed and the primer. In addition:

"A" tube contains all four dNTPs and ddATP,
"G" tube contains all four dNTPs and ddGTP,
"T" tube contains all four dNTPs and ddTTP, and
"C" tube contains all four dNTPs and ddCTP.

As shown earlier, all of the tubes contain a different dideoxy compound present at about one-hundredth the concentration of the normal triphosphate. As the DNA is synthesized, nucleotides are added on to the growing chain by the DNA polymerase. However, on occasion a dideoxyribonucleotide is incorporated into the chain in place of a normal nucleotide, which results in a chain-terminating event. After the reaction is completed in each of the tubes, the newly synthesized DNA is separated according to size. *The terminal nucleotide of each piece of the synthesized DNA corresponds to the dideoxy compound present in the tube.*

To make the method clear, consider the following example of a 25 nucleotide long fragment of DNA, numbered 1–25:

$C_1A_2C_3G_4A_5T_6T_7G_8A_9C_{10}G_{11}A_{12}T_{13}C_{14}G_{15}A_{16}T_{17}C_{18}C_{19}A_{20}G_{21}G_{22}A_{23}$ $C_{24}T_{25}$

The newly synthesized DNA will have the following complimentary sequence of bases:

$G_1T_2G_3C_4T_5A_6A_7C_8T_9G_{10}C_{11}T_{12}A_{13}G_{14}C_{15}T_{16}A_{17}G_{18}G_{19}T_{20}C_{21}C_{22}T_{23}$ $G_{24}A_{25}$

The sequence of the original DNA can be derived by reading the end letters from the shortest to the longest pieces of the newly synthesized DNA. In the earlier example, the following sizes of DNA fragments were found in each tube (each number corresponds to the number of nucleotides in the newly synthesized piece):

"A" tube: 6, 7, 13, 17, and 25;
"G" tube: 1, 3, 10, 14, 18, 19, and 24;
"T" tube: 5, 9, 12, 16, 20, and 23; and
"C" tube: 4, 8, 11, 15, 21, and 22.

Since the Sanger Method of DNA sequencing can only be used for fairly short strands (200–900 nucleotides), longer sequences must be subdivided into smaller fragments, and subsequently reassembled to give the overall sequence. The principal method used for this is "shotgun sequencing," which uses random fragments. DNA is broken up randomly into numerous small segments, each of which is sequenced. Since the breaks in DNA occur at different places along the DNA chain, the sequences that are obtained will overlap. Computer programs then use the overlapping ends of different sequences to assemble them into a continuous sequence. For example, consider the following two rounds of shotgun reads:

First shotgun sequence (45 bases): $A_1G_2C_3A_4T_5G_6C_7T_8G_9C_{10}A_{11}G_{12}T_{13}C_{14}$ $A_{15}T_{16}G_{17}C_{18}T_{19}T_{20}$ $A_{21}G_{22}G_{23}$ $C_{24}T_{25}A_{26}A_{27}G_{28}C_{29}A_{30}T_{31}G_{32}C_{33}T_{34}\mathbf{G_{35}}$ $\mathbf{C_{36}A_{37}G_{38}T_{39}C_{40}A_{41}T_{42}G_{43}C_{44}T_{45}}$
Second shotgun sequence (44 bases): $\mathbf{G_1C_2A_3G_4T_5C_6A_7T_8G_9C_{10}T_{11}}A_{12}G_{13}$ $C_{14}A_{15}$ $T_{16}G_{17}$ $C_{18}C_{19}A_{20}T_{21}G_{22}$ $C_{23}A_{24}T_{25}T_{26}A_{27}T_{28}A_{29}C_{30}C_{31}G_{32}C_{33}C_{34}$ $A_{35}T_{36}C_{37}G_{38}G_{39}A_{40}T_{41}C_{42}A_{43}G_{44}$

As can be seen, bases 35–45 in the first shotgun sequence overlap (are identical) to bases 1–11 in the second shotgun sequence. Thus, eliminating the overlap and merging the two sequences yields the following sequence of 80 bases:

Reconstructed sequence: $A_1G_2C_3A_4T_5G_6C_7T_8G_9C_{10}A_{11}G_{12}T_{13}C_{14}A_{15}T_{16}$ $G_{17}C_{18}T_{19}T_{20}A_{21}G_{22}G_{23}C_{24}T_{25}A_{26}A_{27}G_{28}C_{29}A_{30}T_{31}AG_{32}C_{33}T_{34}G_{35}C_{36}A_{37}$ $G_{38}T_{39}C_{40}A_{41}T_{42}G_{43}C_{44}T_{45}$ $C_{46}T_{47}A_{48}G_{49}$ $C_{50}A_{51}T_{52}G_{53}C_{54}C_{55}A_{56}T_{57}G_{58}$ $C_{59}A_{60}T_{61}T_{62}A_{63}T_{64}A_{65}C_{66}C_{67}G_{68}C_{69}C_{70}A_{71}T_{72}C_{73}G_{74}G_{75}A_{76}$ $T_{77}C_{78}A_{79}G_{80}$

Since the time of the development of the Sanger Method and the shotgun method, high demand for low-cost DNA sequencing has driven the development of automated and computerized sequencing technologies, referred to as "next-generation sequencing." At present, millions of sequences can be obtained concurrently, greatly lowering the cost. For example, the cost to sequence a single human genome has been reduced from $100,000,000 in 2000 to $1,000 in 2015.

For his invention of DNA sequencing methods, Sanger shared the 1980 Nobel Prize in Chemistry with Walter Gilbert of Harvard University and Paul Berg of Stanford University. Sanger has the rare distinction of winning the Nobel Prize twice (in 1958 and 1980), placing him in the company of Marie Curie, Linus Pauling, and John Bardeen. Sanger retired in 1983 and devoted himself to his garden in the Cambridge fens. He refused a knighthood out of principle but accepted the Order of Merit (OM) in 1986 because it does not bestow a title. By all accounts, Sanger was a true "gentle" man, extremely courteous and charming. He died in November 2013 at the age of 95 (Fig. 8.1).

The second very important method developed for DNA research was the PCR technique. PCR stands for *P*olymerase *C*hain *R*eaction. The basic idea of the PCR method is to take a very small amount of DNA and make billions of identical copies of it. The inventor of the PCR technique, Kary Mullis, who at the time was working at Cetus Corporation in California, wrote subsequently in his autobiography the following colorful account of his inspiring discovery and the difficulty he initially had in publishing it:

PCR was a chemical procedure that would make the structures of the molecules of our genes as easy to see as billboards in the desert and as easy to manipulate as Tinker toys. PCR would not require expensive equipment, and it would find tiny fragments of DNA and multiply them billions of times. And it would do it quickly. The procedure would be valuable in diagnosing genetic diseases by looking into a person's genes. It would find infectious diseases by detecting the genes of pathogens that were difficult or impossible to culture. PCR would solve murders from DNA samples in trace materials—semen, blood, hair. The field of molecular paleobiology would blossom because of PCR. Its practitioners would inquire into the specifics of evolution from the DNA in ancient specimens. The branchings and migrations of early man would be revealed from fossil DNA and its descendant DNA in modern humans. And when DNA was finally found on other planets, it would be PCR that would tell us whether we had been there before or whether life

FIGURE 8.1 Photograph of Fredrick Sanger (1918–2013). *(From ®© The Nobel Foundation.)*

on other planets was unrelated to us and had its own separate roots. I knew that PCR would spread across the world like wildfire. This time there was no doubt in my mind: Nature would publish it. They rejected it. So did Science, the second most prestigious journal in the world. Science offered that perhaps my paper could be published in some secondary journal, as they felt it would not be suitable to the needs of their readers. "Fuck them," I said. It was some time before my disgust with the journals mellowed. I accepted an offer by Ray Wu to publish it in Methods in Enzymology, a volume he was preparing. He understood the power of PCR. This experience taught me a thing or two, and I grew up some more. No wise men sit up there, watching the world from the vantage point of their last twenty years of life, making sure that the wisdom they have accumulated is being used.

We have to make it on the basis of our own wit. We have to be aware—when someone comes on the seven o'clock news with word that the global temperature is going up or that the oceans are turning into cesspools or that half the matter is going backward—that the media are at the mercy of the scientists who have the ability to summon them and that those who have such ability are not often minding the store. More likely they are minding their own livelihoods.

Although there is considerable truth in Mullis's statement, most scientists would agree that there are always some open-minded editors and reviewers of scientific manuscripts willing to consider new ideas. More significantly, important discoveries that can be repeated, even if they are initially published in minor journals, eventually are accepted and become part of the foundation of science.

Mullis first presented his basic idea of PCR and some preliminary results on the method in a closed meeting of the Cetus Corporation in 1984. As is the normal industrial practice, the results were used to apply for patents before any press release, publication or public lecture. According to patent law, if a discovery is in the public domain prior to the time of the patent application, the patent application will be rejected. In 1985, Cetus Corporation applied for a patent covering the basic idea of PCR and many potential applications, indicating Kary Mullis as the discoverer. The first scientific paper on PCR appeared on December 20, 1985. After modification, the patent was approved on July 28, 1987 (Mullis K.B. "Process for amplifying nucleic acid sequences," US Patent 4,683,202).

How does PCR work? To begin, the DNA that you want to copy is heated to around 95°C to separate the two strands of DNA. The heating step is necessary because, as you remember, DNA polymerase only copies single-stranded DNA. The DNA solution is then cooled and a mixture of the four deoxyribonucleoside triphosphates, dAppp, dGppp, dTppp, and dCppp), DNA polymerase and a "DNA primer" are added. The process of heating, cooling, and DNA polymerase reaction are repeated several times to obtain sequentially 2, 4, 8, 16, 32, 64, etc. copies of the desired DNA. After 30 reaction cycles more than a billion identical copies are produced from a single copy of a DNA molecule.

One of the problems that Kary Mullis had to overcome is that the DNA polymerase 1 that Arthur Kornberg had purified from the bacterium *Escherichia coli* becomes inactive when heated to 95°C. Thus, after each cycle, he had to add fresh enzyme. This was tedious and expensive. Mullis overcame this problem by using a DNA polymerase purified from a thermophilic (heat-loving) bacterium, *TThermus aquaticus*. This bacterium was originally isolated in 1969 from a hot spring in Yellowstone National Park. In 1976, the DNA polymerase of *T. aquaticus* was isolated and shown to be stable at high temperatures retaining its activity even after being heated to 95°C. Thus, the enzyme would only need to be added once, making the technique less labor intensive and more affordable.

In 1988, a programmable system to perform PCR was introduced. The automated system performs the cyclic temperature changes required for enzymatic amplification of DNA segments in vitro. A microcomputer system controls the temperature flow in a 24-well sample holder so that the temperature of the samples in the holder varies as required for DNA strand separation (94–96°C), primer binding, and DNA polymerization (72°C) activity. The microcomputer automatically performs multiple thermal cycles and is sufficiently flexible that the temperature profile can be varied from cycle to cycle. Today, the PCR

FIGURE 8.2 Photograph of Kary Mullis (1944–). *(Taken from https://en.wikipedia.org/wiki/Kary_Mullis#/media/File:Kary_Mullis.jpg.)*

thermal cycler machine costs about $500 and is available in most molecular biology laboratories.

Kary Mullis was awarded the Nobel Prize in Chemistry in 1993 for his development of PCR. Since winning the Nobel Prize, Mullis has been criticized for promoting ideas in areas in which he has no expertise. He has expressed disagreement with the scientific consensus on human-caused global warming, ozone depletion, the evidence that HIV causes AIDS, and expresses belief in astrology (Fig. 8.2).

The Sanger DNA sequencing method together with the Mullis PCR technology have transformed the field of genetics from a science of descriptive analysis into today's powerful technology of gene analysis, genetic manipulation, and gene therapy. Millions of PCRs are run every day in thousands of laboratories around the world and the DNA amplified by PCR is then sequenced by the Sanger or next-generation sequencing methods.

In the Prologue to this book, I described how a man convicted of a crime that he did not commit was released after 35 years in jail because of DNA evidence. DNA fingerprinting only became possible after DNA sequencing and PCR technology was developed. The combination of PCR and gene sequencing is now widely used in many applications, including medicine, infectious disease, ecology, forensics, and research.

In *medicine*, PCR and gene sequencing technologies are used for genetic testing, where a sample of DNA is analyzed for the presence of mutations that are associated with specific diseases. Thus, prospective parents can be tested for being genetic carriers, or their children might be tested for actually being affected by a disease. DNA samples for prenatal testing can be obtained by amniocentesis, chorionic villus sampling, or even by the analysis of rare fetal cells circulating in the mother's bloodstream. The combined PCR and gene sequencing technologies can also be used as part of a sensitive test for tissue typing, vital to organ transplantation, and to replace the traditional antibody-based tests for blood type. Many forms of cancer involve alterations (mutations) of genes into what is called oncogenes (discussed in Chapter 16). Genetic testing can reveal those mutations responsible for oncogenic variants or for hereditary disease, allowing a medical professional to make informed decisions about treatment, and helping prospective parents know whether their children will be at risk for developing cancer or other diseases. By analyzing the sequence of these mutations, therapeutic regimens can sometimes be individually customized to a patient, be he/she a new born baby or a cancer patient.

PCR/gene sequencing has revolutionized the detection of *infectious disease* organisms. For example, tests have been developed that can detect a single HIV virus genome among the DNA of 50,000 host cells. This allows for early detection of infection, including in newborns, rapid and efficient screening of donated blood and it also allows for the examination of the effects of antiviral treatment. Some disease organisms, such as the causative agent of tuberculosis, *Mycobacterium tuberculosis*, are difficult to culture from patients and grow very slowly in the laboratory. PCR-based tests have allowed rapid detection of small numbers of these and other disease-causing microbes. Detailed genetic analysis can also be used to detect antibiotic resistant bacteria, allowing immediate and alternative effective therapies. The spread of disease organisms through populations of domestic or wild animals can also be monitored by PCR/gene sequencing.

The recent union of molecular genetic methods and *ecology* has also brought a great advance to evolutionary biology research. Molecular ecologists employ an array of molecular tools to study the genetic biodiversity of Earth. In ecology, the initial task is often obtaining a survey of the organisms present in the study site. Most molecular-based ecology studies begin with the extraction of DNA from a particular organism, followed by PCR. The utility of PCR lies in the fact that only minute quantities of DNA are needed. This is particularly useful when researchers are unable to obtain large amounts of tissue (e.g., as in the case of rare and extinct plant or animal species) or when numerous samples are needed, as in the case of population genetic studies. DNA sequencing is then used for determining the evolutionary history of a group of organisms and for inferring evolutionary processes and patterns, historical patterns of migration and expansion of animal and plant species, and the evolution of specific genetic traits.

The development of DNA fingerprinting protocols has seen widespread application in *forensics*. Genetic fingerprinting can uniquely discriminate any one person from the entire population of the world. Minute samples of DNA can be isolated from a crime scene, amplified by PCR, sequenced, and then compared to that from suspects, or from a DNA database of earlier evidence or convicts. Simpler versions of these tests are often used to rapidly rule out suspects during a criminal investigation. Evidence from decades-old crimes can be tested, confirming or exonerating the people originally convicted. Also, DNA fingerprinting can help in parental testing, where an individual is matched with their close relatives. This test is now routinely used, for example, to confirm the biological parents of an adopted (or kidnapped) child.

Since the 1980s, PCR/DNA sequencing has become the foundation for most DNA *research*. The remaining chapters will tell us the fascinating stories of how these technologies have been crucial to discoveries regarding the human genome, the human microbiota (microorganisms associated with humans), and the development of genetic engineering, in addition to detecting and treating cancers, learning about the aging process, and the role of DNA in the evolution of animals, including humans.

NOTES AND REFERENCES

61 Sanger, F., Nicklen, S., Coulson, A.R., 1977. DNA sequencing with chain-terminating inhibitors. Proc. Natl. Acad. Sci. USA 74, 5463–5467.

61 Wright, P., 2013. Frederick Sanger obituary: Nobel prizewinning biochemist whose pioneering work on insulin and DNA transformed the field of genetics. The Guardian, November 20, 2013.

62 Metzenberg, S., 2003. Sanger Method—Dideoxynucleotide Chain Termination. Available from: http://www.csun.edu/~hcbio027/biotechnology/lec3/sanger.html

64 Liu, L., Li, Y., Li, S., et al., 2012. Comparison of next-generation sequencing systems. J. Biomed. Biotechnol. 2012, 1–11.

64 Mullis, K., 2000. Dancing Naked in the Mind Field. Vintage Books, New York, NY.

66 Bartlett, J.M.S., Stirling, D., 2003. A short history of the polymerase chain reaction. PCR Protocols 226, 3–6.

66 Brock, T.D., Freeze, H., 1969. *Thermus aquaticus*, a nonsporulating extreme thermophile. J. Bacteriol. 98, 289–297.

66 Chien, A., Edgar, D.B., Trela, J.M., 1976a. Deoxyribonucleic acid polymerase from the extreme thermophile *Thermus aquaticus*. J. Bacteriol. 174, 1550–1557.

66 Cetus Corporation in Emeryville, California was one of the first biotechnology companies.

66 Chien, A., Edgar, D.B., Trela, J.M., 1976b. Deoxyribonucleic acid polymerase from the extreme thermophile *Thermus aquaticus*. J. Bacteriol. 127, 1550–1557.

67 Two different primers are necessary for PCR, one for each end of the molecule to be copied, because the polymerase starts replication only at the 3′-end of the primer, and the two strands are antiparallel.

67 Weier, H.U., Gray, J.W., 1988. A programmable system to perform the polymerase chain reaction. DNA 7, 441–447.

Johnson, G. Bright Scientists, Dim Notions. The New York Times, August 6, 2010.

68 Quill, E., 2008. Blood-matching goes genetic. Science Magazine, pp. 1478–1479.

69 Kwok, S., et al., 1987. Identification of HIV sequences by using in vitro enzymatic amplification and oligomer cleavage detection. J. Virol. 61, 1690–1694.

Chapter 9

Jumping Genes

It might seem unfair to reward a person for having so much pleasure over the years, asking the maize plant to solve specific problems and then watching its responses.

—Barbara McClintock, commenting on receiving the Nobel Prize

Barbara McClintock's burning curiosity, enthusiasm, and uncompromising honesty serve as a constant reminder of what drew us all to science in the first place.

—Gerald Ralph Fink

It is now my pleasure to introduce one of the great ladies of genetics/DNA research, Barbara McClintock. McClintock's research with the maize (corn) plant was revolutionary in that it showed for the first time that an organism's genome is not a stationary entity, but rather is subject to alteration and rearrangement. This concept of "jumping genes" or transposable elements, or transposons for short, was met initially with criticism from the scientific community. However, the role of transposons eventually became widely appreciated, and McClintock was awarded the Nobel Prize in 1983 in recognition "for her discovery of mobile genetic elements." Her discoveries have had a profound effect on everything from evolutionary biology to cancer research and also opened the door to genetic engineering.

Barbara McClintock was born in 1902 in Connecticut to Thomas and Sara McClintock, the third of four children. The McClintock family moved to Brooklyn in 1908, and Barbara completed her secondary education there at Erasmus Hall High School, graduating in 1919. She was an independent child beginning at a very young age, a trait she later identified as her "capacity to be alone." In high school, McClintock discovered her love of science and reaffirmed her solitary personality. She wanted to continue her studies at Cornell University's College of Agriculture. However, her mother resisted sending McClintock to college, for fear that she would be unmarriageable, but her father intervened and she began her studies at Cornell in 1919.

At Cornell, she participated in student government and was invited to join a sorority, though she soon realized that she preferred not to join formal organizations. McClintock studied plant breeding and botany, receiving a BSc in 1923 in Agriculture. Her interest in genetics began when she took her first course in that field in 1921. The course was given by C.B. Hutchison, a plant breeder and

It's in Your DNA. http://dx.doi.org/10.1016/B978-0-12-812502-1.00009-3

geneticist. Hutchison was impressed by McClintock's interest and intelligence, and invited her to participate in the graduate genetics course at Cornell in 1922. McClintock pointed to Hutchison's invitation as the reason she continued in genetics: "Obviously, this invitation cast the die for my future. I remained with genetics thereafter." McClintock received both her master's (1925) and doctoral degrees (1927) from Cornell's College of Agriculture. After completing requirements for the PhD degree in the spring of 1927, she remained at Cornell to initiate studies aimed at associating each of the 10 chromosomes comprising the maize complement with the genes each carries.

During her graduate and postgraduate studies at Cornell, McClintock's maize research focused on cytology, the branch of life science which deals with the study of cell structures, in her case, developing ways to visualize and characterize maize chromosomes under the microscope. She was the first to describe the sizes and detailed shapes of the 10 maize chromosomes. By combining her microscopic studies of the morphology of chromosomes with genetic traits of maize, McClintock pioneered the field of cytogenetics. In 1931, McClintock published the first genetic map for maize, showing the order of genes on maize chromosome number 9. The following year, she observed how the recombination (crossing over) of chromosomes seen under a microscope correlated with new traits. Crossing over is the swapping of genetic material that occurs in the germ line during sexual division. Paired chromosomes from each parent align so that similar DNA sequences from the paired chromosomes cross over one another. Crossing over results in a shuffling of genetic material and is an important cause of the genetic variation seen among offspring. (This is the main reason why children of the same parents are genetically different.) Further, McClintock showed that genetic traits that were linked, in other words, traits that were always transferred together to progeny, were present next to each other on the same chromosome. These important findings led to her being awarded several postdoctoral fellowships, but she was not offered an academic position at Cornell. McClintock received a fellowship from the Guggenheim Foundation that made possible research in Germany during 1934. However, she left Germany early amidst mounting political tension in Europe, and returned to Cornell, remaining there until 1936, when she accepted an Assistant Professorship offered to her by the Department of Botany at the University of Missouri, located in Columbia, Missouri.

During her time at Missouri, McClintock worked with the geneticist Lewis Stadler, who introduced her to the use of X-rays as a mutagen. Exposure to X-rays increases the rate of mutation above the natural background level, making it a powerful research tool for genetics. Through her work with X-ray-mutagenized maize, she identified ring chromosomes, which form when the ends of a single chromosome fuse together after radiation damage. She showed that the subsequent loss of ring-chromosomes caused mass mutations observed as different colors in irregular patches or streaks (variegated) in maize seeds. This was a key cytogenetic discovery for several reasons. First, it showed that the rejoining of chromosomes was not a random event, and second, it demonstrated a source of large-scale mutation. As I will discuss in Chapter 16, these types of chromosomal rearrangements are closely linked to the development of cancer. For this reason, cytogenetics remains an area of interest in cancer research today.

Although her research was progressing at Missouri, McClintock was not satisfied with her position at the University. She was not promoted, excluded from faculty meetings, and was not made aware of positions available at other institutions. In 1940, she wrote to a former colleague at Cornell University, Charles Burnham: "I have decided that I must look for another job. As far as I can make out, there is nothing more for me here. I am an assistant professor at $3,000 per year, and I feel sure that that is the limit for me." McClintock, 38 years old with numerous important publications, was denied senior faculty positions at Cornell University and the University of Missouri due to discrimination against women scientists. While being expected to recommend male colleagues to universities like Yale and Harvard, she could not find a research-related position herself. She had lost trust in her employer and in the University administration, and early in 1942, McClintock took a leave of absence from Missouri in hopes of finding a position elsewhere.

McClintock's situation was not unique. Many outstanding female scientists had a difficult time finding academic positions and getting promoted in the United States and Europe throughout the 20th century. In 1973, for example, women made up only 5.7% of the full professors in biology departments at American universities. Since then, the situation has continuously improved, so that today women make up around 28% of full professors in biology departments. This gender gap is the subject of hot debate, as illustrated in 2005, when then-Harvard president Larry Summers argued that differences in science aptitude between men and women explained most of the problem. Although not directly related to McClintock, sociological research has found that the root of the gender gap problem is often the clash between career and child-rearing, especially given that the long road through graduate school, postdoctoral research positions at universities, and tenure-track professorship meanders through a person's 20s and 30s, a time when women are disproportionately more likely than men to have childbearing and child care responsibilities.

In the summer of 1941, Milislav Demerec, a Croatian-American geneticist, invited McClintock to join him at Cold Spring Harbor Laboratory, a private, nonprofit institution in New York that is ranked among the leading basic research institutions in the world in molecular biology and genetics. Having carried out research himself in corn genetics, Demerec knew and respected McClintock as a scientist. McClintock stayed for the summer and late into the fall at the Cold Spring Harbor Laboratory. When Demerec became the Director of the Department of Genetics of the Carnegie Institution of Washington at the Cold Spring Harbor Laboratory, he offered McClintock a research position. Undecided at first, McClintock finally accepted the position in 1942 and remained there until her retirement in 1967.

FIGURE 9.1 **Unstable color patterns of corn seeds.**

McClintock's position at Cold Spring Harbor changed the course of her life and career. It gave McClintock the freedom to pursue her own research without the obligations of teaching or constantly applying for outside funds. The extensive resources of the institution provided her with land on which to cultivate maize. McClintock's position at Cold Spring Harbor also enabled her to avoid much of the scrutiny often imposed upon women scientists. She was free to do as she pleased (Fig. 9.1).

In the summer of 1944, McClintock began systematic studies on the mechanisms of the mosaic color patterns of maize seed, which would lead to her revolutionary discovery of "jumping genes" or what is now referred to as gene transposition. She had observed previously that the mosaic or variegated pattern of the seeds was an unstable property. It changed too frequently from one generation to the next to be explained by simple mutation. What then was responsible for the unstable color patterns of the seeds? Among the plants she grew that summer, she found two new genes that were located on the short arm of chromosome 9. She named these genes "Dissociator" (Ds) and "Activator" (Ac). Contrary to its name, Dissociator gene did not merely dissociate, or break, the chromosome. It turned out to have a variety of effects on neighboring genes, but only when the Activator gene was also present. In early 1948, she made the surprising discovery that both Dissociator and Activator genes could transpose, or change positions on the chromosome. She observed the effects of the transposition of Ac and Ds genes by the changing patterns of coloration in maize kernels over generations of controlled crosses, and described the relationship between the two positions through intricate microscopic analysis. She concluded that Ac controls the transposition of the Ds from chromosome number 9, and that the movement of Ds is accompanied by the breakage of the chromosome. When Ds moves, a gene responsible for producing purple color is transformed into the active form, which initiates the pigment (anthocyanin) synthesis in cells. The transposition of Ds in different cells is random, it may move in some cells but not others, which causes the mosaic color. The size of the purple colored spot on the seed is determined by the stage of the seed development at which the transposition took place. If it took place at an early stage, pigment production occurred for a relatively long time and the spot would be large; if it took place late, the spot would be small.

McClintock's discovery challenged the concept of the genome as a static set of instructions passed between generations. In 1950, she reported her work on Ac/Ds and her ideas about gene regulation in a paper entitled "The origin and behavior of mutable loci in maize" published in the journal *Proceedings of the National Academy of Sciences (USA)*. Her work on controlling elements and gene regulation was conceptually difficult and was not immediately understood or accepted by her contemporaries; she described the reception of her research as "puzzlement, even hostility." Nevertheless, McClintock continued to develop her ideas on controlling elements. She published a paper in the journal *Genetics* in 1953, where she presented all her statistical data. Based on the reactions of other scientists to her work, McClintock felt she risked alienating the scientific mainstream, and from 1953 she stopped publishing accounts of her research on controlling elements.

In 1957, McClintock received funding from the National Academy of Sciences to start research on indigenous strains of maize in Central America and South America. She was interested in studying the evolution of maize through chromosomal changes. McClintock explored the chromosomal, morphological, and evolutionary characteristics of various races of maize. After extensive work in the 1960s and 1970s, McClintock and her collaborators published the seminal study "The Chromosomal Constitution of Races of Maize," leaving their mark on evolutionary biology, paleobotany (dealing with the recovery and identification of plant remains from geological contexts), and ethnobotany (the study of the relationships that exist between peoples and plants).

McClintock received recognition and many honors in the later part of her life. Most notably, she received the Nobel Prize for Physiology or Medicine in 1983, the first woman to win that prize unshared, credited by the Nobel Foundation for discovering "mobile genetic elements." McClintock was 81 years old when she received the Nobel Prize and it was more than 30 years after she initially described the phenomenon of controlling elements. She was compared to Gregor Mendel in terms of her scientific career by the Swedish Academy of Sciences when she was awarded the Prize. McClintock wrote: "One must await the right time for conceptual change."

McClintock's status as role model of women was significant enough that, in 1947, the American Association of University Women gave her an award.

"The award ceremony was unexpectedly pleasant," McClintock recalled afterward. "It was the handsomest group of women I have ever seen—handsome as only intelligent, middle-aged women can be."

McClintock was never a professor, but she was always interested in teaching young students. When she visited my University in the 1990s, she chose to sit around a table and talk to graduate students, rather than give a lecture or speak to professors. Many people regarded her as an eccentric, but she was just in love with science; that was the main thing in her life (Figs. 9.2 and 9.3).

FIGURE 9.2 Barbara McClintock at her microscope.

FIGURE 9.3 McClintock's microscope and ears of corn on exhibition at the National Museum of Natural History in Washington, DC.

McClintock's pioneering discovery of mobile genes was widely appreciated only after other researchers finally discovered the process of transposition in bacteria in the 1970s. One of the stories of mobile genes in bacteria can be traced back to the sudden development of multiple antibiotic resistances by pathogenic bacteria in the 1950s. At the end of World War II, the antibacterial drug sulfonamide was introduced in Japan for the treatment of dysentery and proved to be effective for the first several years, reducing the incidence of the disease from 100,000 to 15,000. However, after 1950, the incidence again increased, in spite of the fact that sulfonamide was used extensively. Most of the *Shigella* bacteria responsible for the disease were found to be resistant to the drug. In an attempt to overcome the resistance to sulfonamide, newly discovered antibiotics, such as streptomycin, chloramphenicol, and tetracycline were used for the treatment of sulfonamide-resistant *Shigella*. Again there was initial success with a large decrease in the occurrence of dysentery. However, between 1954 and 1964 the frequency of *Shigella* strains in Japanese hospitals that were simultaneously resistant to two or more of these drugs, referred to as *multiple-resistant strains*, rose from 0.2% to 52%. What was going on?

In 1960s, several Japanese researchers showed that multiple-drug resistance could be transferred between *Shigella* and other bacteria. For example, if multidrug resistant *Shigella* came in contact with *Escherichia coli*, either in a test tube or in the intestines, a high percentage of the *E. coli* strains became multidrug resistant. Furthermore, the researchers found that the genes responsible for the resistance were located on plasmids. Plasmids are relatively small, circular DNA molecules which replicate in the bacterial cytoplasm and are separate from the bacterial chromosome. Such plasmids were termed drug-resistance (R) factors. We now know that R factors are commonly responsible for the sweeping spread of multiple drug-resistant bacteria. Many different disease-causing bacteria have acquired multidrug resistance because of the excessive and often unnecessary use of antibiotics in humans and animals. Today, antibiotic resistance has become a major clinical and public health problem.

Genes responsible for drug resistance can not only be transferred between different bacteria, but they can also be transferred between the plasmid and the bacterial chromosome. The mobility of antibiotic-resistant genes is only one example of the general phenomenon of *horizontal gene transfer* (HGT). This phenomenon includes the movement of genetic material within a single cell or bacterium to different places on the chromosome and between different organisms. This genetic transfer differs from the traditional "vertical" transfer of genes from the parental generation to offspring. HGT is the primary reason for bacterial antibiotic resistance and is precisely the same concept of "jumping genes" that McClintock reported in maize in 1950.

The manner in which HGT takes place has been studied in bacteria, animals, and plants. The basic finding is that genes that move in and out of genomes by a "cut and paste" mechanism are present on transposable elements, referred to as transposons. A transposon is a segment of DNA of varying length that has two special properties that allow for HGT. (1) The ends of the transposon have a short-inverted repeat of bases; for example, if one end of the transposon has the sequence AGGTCAGATG, the other end would have the sequence in the opposite direction, GTAGACTGGA. (2) Inside the transposon is a gene coding for the enzyme transposase. This enzyme recognizes the short sequences at the ends of the transposon, makes breaks at those locations (excision), and reinserts the transposon at another location.

Transposons can cause mutations, change the amount of DNA in the cell and dramatically influence the structure and function of the genomes where they reside. The movement of genes that are on transposons can cause diseases that include hemophilia, severe combined immunodeficiency, porphyria, muscular dystrophy, and predisposition to cancer. The connection between mutation, cancer, and transposons are discussed in Chapter 16. Transposons in bacteria often carry additional genes for functions other than transposition—often for antibiotic resistance. In bacteria, transposons can jump from chromosomal DNA to plasmid DNA and back, allowing for the transfer and permanent addition of genes, such as those encoding antibiotic resistance (multiantibiotic-resistant bacterial strains can be generated in this way).

Transposons are astonishingly abundant, comprising a majority of the DNA in some species. Transposons comprise more than 65% of the human genome and approximately 85% of the maize genome. We now know that the genes that McClintock named "Activator" (Ac) and "Dissociator" (Ds) are transposons. Ac is a complete transposon that can produce a functional transposase, which is required for the element to move within the genome. Ds has a mutation in its transposase gene, which means that it cannot move without another source of transposase. Thus, as McClintock observed, Ds cannot move in the absence of Ac.

The pioneering discovery of jumping genes by McClintock and the subsequent confirmation and elaboration of her concept of horizontal gene transfer (HGT) brought about a paradigm change in genetics and evolutionary biology. Previously, the entire emphasis was on the role of point mutations in coding regions as the primary source of evolutionary change. A point mutation is the replacement of one base by another in DNA. For example, a mutation in the codon ACC to yield GCC would lead to the replacement of the amino acid threonine by alanine in a particular protein. Since many such point mutations would be required to construct a new enzyme and point mutations are random and infrequent, it would take an extremely long time for an organism to evolve a gene that codes for a new enzyme. HGT, on the other hand, can introduce one or more novel genes in a single step.

Recent DNA analyses of the genomes of different animals and plants revealed that HGT has played a major role in many evolutionary leaps. One particularly interesting example of HGT in evolution is the gene coding for the protein syncytin in mammals which is required for the development of the placenta. Placental mammals include such diverse forms as bats, whales, elephants, shrews, and some of the most familiar organisms to us, including pets, such as dogs and cats, as well as many farm animals, such as sheep, cattle, and horses. Humans, of course, are also placental mammals. The syncytin gene was transferred into the mammal genome from a virus DNA by HGT. The protein syncytin originally allowed viruses to fuse host cells together which enabled spreading of the virus from one cell to another. Now the viral protein allows babies to fuse to their mothers. Research on syncytin genes has shown that HGT of viral genes into host genomes led to a major evolutionary leap, the formation of placental mammals.

In addition to being an important factor in the evolution of all organisms, HGT is the basis of artificial genetic modification engineering, also called genetic engineering, which is the subject of the next chapter.

NOTES AND REFERENCES

71 Kass, L.B., Provine, W.B., 1997. Genetics in the roaring 20s: the influence of Cornell's professors and curriculum on Barbara McClintock's development as a cytogeneticist. Am. J. Botany 84 (6 Suppl.), 123.

72 Education and Research at Cornell, 1925–1931, the Barbara McClintock Papers, Profiles in Science (National Library of Medicine).

72 Breakage-Fusion-Bridge: The University of Missouri, 1936–1941, the Barbara McClintock Papers, Profiles in Science (National Library of Medicine).

73 Kass, L.B., 2005. Missouri compromise: tenure or freedom. New evidence clarifies why Barbara McClintock left Academe. Maize Genet. Coop. News Lett. 79, 52–71.

75 McClintock, B., 1950. The origin and behavior of mutable loci in maize. Proc. Natl. Acad. Sci. USA 36, 344–355.

Controlling Elements: Cold Spring Harbor, 1942–1967, the Barbara McClintock Papers, Profiles in Science (National Library of Medicine).

75 Searching for the Origins of Maize in South America, 1957–1981, the Barbara McClintock Papers, Profiles in Science (National Library of Medicine).

76 The Louisa Gross Horwitz Prize for Biology or Biochemistry, cumc.columbia.edu (Columbia University).

76 Kolata, G., 1992. Dr. Barbara McClintock, 90, Gene Research Pioneer, Dies, The New York Times.

78 Gyles, C., Boerlin, P., 2014. Horizontally transferred genetic elements and their role in pathogenesis of bacterial disease. Vet. Pathol. 51, 328–340.

78 Reznikoff, W.S., 2003. Tn5 as a model for understanding DNA transposition. Mol. Microbiol. 47, 1199–2006.

79 Dupressoir, A., Lavialle, C., Heidmann, T., 2012. From ancestral infectious retroviruses to bona fide cellular genes: role of the captured syncytins in placentation. Placenta 33, 663–671.

Chapter 10

Genetic Engineering

I suspect any worries about genetic engineering may be unnecessary. Genetic mutations have always happened naturally, anyway.

—James Lovelock, futurist and member of the Royal Society of London

Recombinant DNA technology [genetic engineering] faces our society with problems unprecedented not only in the history of science, but of life on the Earth. It places in human hands the capacity to redesign living organisms, the products of some three billion years of evolution. The nub of the new technology is to move genes back and forth, not only across species lines, but across any boundaries that now divide living organisms. The results will be essentially new organisms, self-perpetuating and hence permanent. Once created, they cannot be recalled. It presents probably the largest ethical problem that science has ever had to face. Our morality up to now has been to go ahead without restriction to learn all that we can about nature. Restructuring nature was not part of the bargain. For going ahead in this direction may be not only unwise, but dangerous. Potentially, it could breed new animal and plant diseases, new sources of cancer, novel epidemics.

—George Wald, Nobel Laureate in Medicine

The time to talk about it (genetic engineering to improve a baby's genes) in schools and churches and magazines and debate societies is now. If you wait, 5 years from now the gene doctor will be hanging out the MAKE A SMARTER BABY sign down the street.

—Arthur L. Caplan, Professor and Head of the Division of Bioethics at New York University

By the early 1970s, it became clear to DNA researchers that recent advances in techniques made it possible to perform genetic engineering on live organisms. Genetic engineering, also called recombinant DNA technology, involves the group of techniques used to cut up and join together genetic material, especially DNA from different biological species, and to introduce the resulting hybrid DNA into an organism in order to form new combinations of heritable genetic material and thereby change one or more of its characteristics. The main difference between genetic engineering and natural genetic variations is that in genetic engineering the different pieces of DNA used for the recombination can come from any organism even those that are far apart as a bacterium and a cow or plant.

It's in Your DNA. http://dx.doi.org/10.1016/B978-0-12-812502-1.00010-X

In 1972, Paul Berg at Stanford University created the first genetically engineered virus by combining DNA from a monkey virus with that of a bacterial virus. In 1973, Herbert Boyer at University of California at San Francisco and Stanley Cohen at Stanford University created the first genetically engineered bacterium by inserting DNA coding for antibiotic resistance into a plasmid of *Escherichia coli*. A year later, Rudolf Jaenisch, while he was a postdoc at Princeton University, created the world's first genetically engineered animal by introducing foreign viral DNA into the genome of a mouse embryo.

These achievements led to concerns in the scientific community about potential risks from genetic engineering. An ad hoc committee was appointed by the National Academy of Sciences, USA, in 1974 to assess the dangers of genetic manipulation. Paul Berg headed the committee, after he had halted his own genetic engineering experiments and had campaigned for a moratorium on such research pending the establishment of a framework for ensuring its safety. The Berg Committee concluded that an international conference was necessary to resolve the issue and that until that time scientists should halt experiments involving genetic engineering. This moratorium on genetic engineering research was the first such voluntary moratorium in the history of science.

The International Conference that Berg helped to organize was held on the 24th–27th of February, at the Asilomar Conference Center in Pacific Grove, California. The Conference was attended by 90 American scientists, 60 scientists from 12 other countries, physicians, lawyers, and 16 members of the press. The main focus of the Conference was to draw up voluntary guidelines to ensure the safety of genetic engineering research for the laboratory scientists working in this field as well as the general public. The major concern was that some of the artificial recombinant DNA molecules could prove biologically hazardous. For example, in most genetic engineering experiments, the recombinant molecule is first introduced into the bacterium *E. coli*. The bacterium is then grown for several hours to increase the number of recombinant molecules. Strains of *E. coli* are common residents of the intestinal tract of humans and domestic animals, and they are capable of exchanging genetic material with other bacteria, some of which can be pathogens. Thus, recombinant DNA molecules introduced into *E. coli* might possibly be disseminated among human, bacterial, plant, or animal populations with unpredictable consequences.

While acknowledging that they could not predict or avert hazards with any certainty, the scientists at the Asilomar Conference classified each type of genetic engineering experiment according to its perceived risk to laboratory personnel and public safety. They determined that some low-risk experiments should proceed, while others required strict containment or even temporary prohibition. Researchers were required to wear gloves and masks and perform the experiments in negative pressure hoods that prevent air from escaping into the laboratory. In addition to physical barriers, the Conference advocated the use of bacteria that were unable to survive in natural environments and plasmids and viruses that were able to multiple only in specified hosts. The voluntary

FIGURE 10.1 Four of the major participants at the Asilomar Conference. Left to right: Maxine Singer, Norton Zinder, Sydney Brenner, Paul Berg.

guidelines presented at the Asilomar Conference was replaced in June 1976 by a formal set of "Guidelines for Research Involving Recombinant DNA Molecules" established by the National Institutes of Health (NIH). These guidelines are periodically updated as more information becomes available (Fig. 10.1).

The Asilomar Conference was a milestone in the development of social awareness and of public responsibility among scientists. Faced with real questions of theoretical risks, the scientists paused and then decided to proceed with caution. Uncertainty of risk is a compelling reason for caution. It will occur again in some areas of scientific research, and the initial response must be the same.

It should be emphasized that the Asilomar and NIH guidelines dealt exclusively with the safety of laboratory experiments. No consideration at that time was given to the ecological safety, economics, and ethics of actually commercially producing genetically engineered products, now commonly referred to as genetically modified (GM) products.

When the ban on performing genetic engineering experiments was abolished, a few of the innovators of the new technology realized its commercial potential and established private biotechnology companies. One of first to do this was Herbert Boyer who together with a 29-year-old venture capitalist Robert Swanson founded Genentech Inc. in South San Francisco, California. As their first genetic engineering project, the company chose to develop the production of human insulin in bacteria. Insulin is a protein hormone that is produced in the pancreas and that controls the metabolism of glucose and other carbohydrates.

In September 1978, Genentech announced the successful production of human insulin in bacteria using recombinant DNA technology.

In the United States, approximately 20 million diabetics take injections of insulin. Before the development of genetically engineered insulin, the hormone was extracted from the pancreas glands of swine and cattle. It took about 8000 lbs of animal pancreas glands to produce 1 lb of insulin. The major advantage of producing insulin using the recombinant DNA method is that the insulin is identical to human insulin. Previously, many diabetics had suffered allergic reactions to insulin derived from animals. Other advantages of genetically engineered insulin are much lower production costs, overcoming the ethical objection to sacrificing animals, and the much smaller doses needed for the treatment. In 1982, human insulin of recombinant DNA origin was approved by the appropriate drug regulatory agencies in the United Kingdom, the Netherlands, West Germany, and the United States. This new source has provided a reliable, expandable, and constant supply of human insulin for diabetics around the world. In 2014, the market for genetically engineered insulin in the United States reached 22 billion USD.

Without getting into the experimental details, recombinant proteins, such as insulin, are produced in microorganisms by the following steps:

1. The specific DNA molecule that codes for the desired protein is obtained, often synthesized using the polymerase chain reaction (PCR) reaction described in Chapter 8.
2. The specific DNA is then stitched into circular DNA plasmid in vitro using special enzymes to perform the molecular surgery; the plasmid chosen for the construction is one that will multiply and yield many copies in each microbial cell.
3. The plasmid carrying the protein gene is transferred into a microorganism, usually *E. coli* or yeast.
4. The genetically engineered microbe is then grown in large fermentation tanks; while inside the microbe, the plasmid gene coding for the desired protein is "switched-on" by the microbe to translate the required and incorporated gene into the protein; the process is the same as that used by microbes to produce their own proteins except that multiple copies of the plasmid ensure that large quantities of the recombinant protein are produced.
5. When the microbes produce sufficient amounts of the protein, the fermentation is halted and the desired protein is purified.

In this manner, many recombinant proteins in addition to insulin have been produced commercially: human growth hormones to prevent dwarfism, follistim (follicle stimulating hormone) to treat infertility problems in women and men, blood-clotting factors for treating hemophilia (the "bleeding disease"), human albumin to treat different disease conditions, such as hemorrhagic shock, cirrhosis with ascites, and other critical clinical conditions related to plasma volume, vaccines and many other drugs. By 2010, the US Food and

Drug Administration (FDA) had licensed 160 protein-based recombinant pharmaceuticals.

As an indication of the significance of recombinant DNA technology, Genentech Inc., which now employs 12,300 people, was sold to the Swiss global health-care company F. Hoffmann-La Roche AG in 2009 for approximately 46.8 billion USD. A large part of the commercial success of Genentech as well as other biotech companies was the 1980 decision of the US Supreme Court to allow the patenting of genetically engineered organisms—*A Living, man-made micro-organism is a patentable subject matter as a "manufacture" or "composition of matter." The fact that the organism sought to be patented is alive is no bar to patentability.*

The two major criteria for granting a patent are originality and potential applicability. Until the 1980 decision, living organisms were not patentable because they were natural, not manmade. For example, it was possible to patent a process for the isolation of an antibiotic, but not the microorganism that produces the antibiotic. The Court ruled that genetically engineered organisms are manmade inventions and therefore novel. This ruling was the force that encouraged private investment in research, development, and marketing of recombinant DNA technology because the process of developing the technology for producing a desired recombinant protein is usually long and costly.

Unlike the production of drugs by genetically engineered organisms, the development of genetically modified (GM) food plants and animals, starting in the 1980s, has been highly controversial. I will discuss GM food crops before GM animals because to-date most of the research, development, and commercialization have been carried out on GM plants. Before discussing the pros and cons of GM crops, the two methods that are used to introduce desired genes into plants to generate genetically engineered crops will be discussed—the biolistic (gene gun) method and use of the bacterium *Agrobacterium tumefaciens* as a vector for DNA transformation. These two special methods were designed mainly in order to overcome the difficulty in penetrating the hard plant cell wall.

In the gene gun method, DNA molecules containing the desired gene or genes are bound to tiny particles of gold or tungsten which are subsequently shot into plant tissue or single plant cells under high pressure. The accelerated particles penetrate both the cell wall and membranes. Inside the cell, the DNA separates from the metal and is integrated into plant the genome inside the nucleus. This method has been applied successfully for genetically engineering many cultivated crops, such as wheat and maize.

The second method, using the bacterium *A. tumefaciens* for introducing foreign genes into plants, takes advantage of the unique natural infection mechanism of this plant pathogen. When this bacterium comes in contact with a plant root it transfers a tumor-inducing plasmid (T-DNA) into the genome of the plant. Normally, a few genes encoded by the T-DNA alter plant hormone levels, leading to uncontrolled cell division and plant tumor formation. Genetic engineering of plants is accomplished by cutting out of the plasmid DNA those genes that cause the tumor formation and replacing them with the specific genes that one desires to introduce into the plant. Thus, the bacterium serves as a vector, enabling transportation of foreign genes into plants. This method works especially well for certain plants like potatoes, tomatoes, and tobacco, but is less successful in crops, such as wheat and maize.

What genes have been put into agricultural crops that make them more productive and resistant to disease? One of the most famous groups of GM crops are those that contain a gene from bacteria that make the plant resistant to the herbicide with the trade name "Roundup." Monsanto Company, a publicly traded American multinational agrochemical and agricultural biotechnology corporation headquartered in St. Louis, Missouri, is the leading producer of Roundup and genetically engineered seed. Roundup kills plants by interfering with the synthesis of three amino acids that are essential for the growth of plants, phenylalanine, tyrosine, and tryptophan. The rationale behind genetically engineering Roundup-resistant agricultural plants is that spraying with Roundup kills the weeds but not the crop, thus giving an agricultural edge to the farmer. The bacterial DNA coding for the resistance to Roundup was injected into soybeans using the gene gun method described earlier. In 1990, Monsanto obtained a patent for agricultural crops genetically engineered to be resistant to Roundup. The patent gave Monsanto exclusive rights to produce and sell seeds resistant to Roundup, but it took another 5 years to receive marketing approval from the United States Department of Agriculture (USDA) for soybeans resistant to Roundup. The adoption of Roundup resistant GM crops worldwide has been rapid and impressive, reaching 150 million hectare, or 800 thousand square miles, annually and now including soybean, maize, cotton, canola, and sugar beet (Fig. 10.2).

Many different kinds of GM seeds are now used in agriculture to produce plants with beneficial characteristics. In addition to herbicide resistance, these traits include improved shelf life, disease resistance, stress resistance, and

FIGURE 10.2 Genetically engineered (GM) plants resistant to Roundup flourish when sprayed with the herbicide, while weeds are killed.

resistance to insects. Examples of genetically engineered foods include an apple that has been modified to resist browning, known as the Nonbrowning Arctic Apple, a genetically modified cassava enhanced with protein and other nutrients (called BioCassava), maize genetically modified to resist drought (called DroughtGard), and a seed oil crop, *Camelina sativa*, that has been engineered to accumulate high levels of the beneficial fish oil omega-3 long-chain polyunsaturated fatty acids. In addition, tobacco, corn, rice, and many other crops that have been genetically engineered to express a gene from the bacterium *Bacillus thuringiensis* that encodes a natural protein that kills insects. These latter plants are referred to as Bt corn, Bt rice, etc.

The recombinant DNA crop that probably has the greatest potential to improve human health worldwide is Golden rice, created to fight vitamin A deficiency, which affects 250 million people around the world and can cause blindness and even death. Rice is one of the most common foods on Earth. In fact, almost half of the world's population survives primarily on rice. As getting vitamin A supplements to every single person on the planet would be impractical, scientists believed that the answer was to create a grain of rice that already had vitamin A in it. Golden rice was engineered to produce beta-carotene in the rice grain. The human body converts beta-carotene into vitamin A. The name Golden rice comes from the bright golden glow the added beta-carotene imparts to the rice (Fig. 10.3).

Free licenses for developing countries were granted quickly due to the positive publicity that Golden rice received. The cutoff between humanitarian and commercial use was set at US$10,000. Therefore, as long as a farmer or subsequent user of Golden rice does not make more than $10,000 per year, no royalties need to be paid. In addition, farmers are permitted to keep and replant the patented seed. In 2014 approximately 16 million farmers grew

FIGURE 10.3 Genetically engineered Golden rice (right side) has a great potential to cover micronutrient needs of rural, rice-based societies.

GM crops in 30 countries. Over 90% of these farmers were resource-poor in developing countries.

Before discussing the pros and cons of genetic engineering, I will briefly describe a very recent and powerful technique for modifying the DNA of bacteria, plants, animals, and humans. The technique is referred to as CRISPR, an acronym for *C*lustered *R*egularly *I*nterspaced *S*hort *P*alindromic *R*epeat. This name refers to the unique organization of short, repeated DNA sequences found in the genomes of bacteria, and other microorganisms. CRISPR technology involves short RNA molecules that match a specific DNA sequence, for example, in a human cell. This "guide RNA" brings molecular machinery to the intended DNA target. Once localized to the DNA region of interest, the molecular machinery can silence a gene or even change the sequence of a gene. This type of gene editing can be likened to edit a sentence with a word processor to delete words or correct spelling mistakes. With early successes in the laboratory, many scientists are looking toward medical and agricultural applications of CRISPR technology, and several Biotech start-ups have been launched that plan to use CRISPR technology to treat human diseases.

The heated controversy over the use of foods and other goods derived from genetically modified crops instead of conventional crops has resulted in peaceful and violent protests, numerous lawsuits, over 2000 scientific papers on the pros and cons of GM foods, and governmental actions to allow GM foods, ban them, or allow them with proper labeling. The proadvocacy group includes farmers, agribusiness experts, scientists, academics, medical doctors, nutritionists, and "company experts" from founding members of the Council for Biotechnology Information. Founding members include BASF, Bayer CropScience, Dow AgroSciences, DuPont, Monsanto Company, and Syngenta. Opponents of genetically modified food, such as the Organic Consumers Association, the Union of Concerned Scientists, GM Watch, Truth in Labeling Coalition, and Greenpeace, claim the proadvocacy groups are biased because they stand to gain financially from GM foods, and many of the scientists receive research funds from the Council for Biotechnology Information.

The advantages of using GM seeds in agriculture are clear—higher productivity, low costs, and in some cases improved nutritional value. Furthermore, for Bt crops (insect resistant) there is an environmental advantage because of the reduced need for chemical insecticides. While there is concern among the public that eating genetically modified food may be harmful, there is broad scientific consensus that food on the market derived from these crops poses no greater risk to human health than conventional food. No reports of ill effects have been documented in the human population from genetically modified food.

Most concerns about GM foods fall into three categories: environmental hazards, human health risks, and socioeconomic concerns. One environmental concern is that GM crops will transfer their foreign DNA to other plants, such as weeds. If herbicide resistant genes are transferred to weeds, the resulting "super weeds" would then be herbicide tolerant. Other introduced genes may cross over into nonmodified crops planted next to GM crops. The possibility of interbreeding is highlighted by the defense of farmers against lawsuits filed by Monsanto. The company filed patent infringement lawsuits against farmers who may have harvested GM crops. Monsanto claims that the farmers obtained Monsanto-licensed GM seeds from an unknown source and did not pay royalties to Monsanto. The farmers claim that their unmodified crops were cross-pollinated from someone else's GM crops planted a field or two away.

Another environmental concern is the unintended harm to other organisms in the environment. A laboratory study published in the prestigious journal *Nature* showed that pollen from Bt corn caused high mortality rates in monarch butterfly caterpillars. Although monarch caterpillars, one of the more beautiful creatures on our planet, normally consume milkweed plants, not corn, the fear is that if pollen from Bt corn is blown by the wind onto milkweed plants in neighboring fields, the caterpillars could eat the pollen and perish. Subsequently, it was shown in the field that the effects of Bt maize cultivation on monarch butterfly populations are negligible, and current evidence suggests that Bt maize is an environmentally safer insect control strategy than conventional chemical spraying. Nevertheless, the possible nontarget effects of Bt crops cannot be excluded. Just as some populations of mosquitoes developed resistance to the now-banned pesticide DDT, many people are concerned that insects will become resistant to Bt or other crops that have been genetically modified to produce their own pesticides.

The major human health risk of GM foods is allergenicity. Many children develop life-threatening allergies to peanuts and other foods. There is a possibility that introducing into a plant a new gene encoding a new protein that might act as an allergen may cause an allergic reaction in susceptible individuals. Extensive testing of all GM foods is now required to avoid the possibility of harm to consumers with food allergies. If even a small percentage of people are allergic to GM foods, it will be necessary to label GM foods and food products. It should also be pointed out that genetic engineering can also be used to remove allergenic proteins that are found in natural foods.

Socioeconomic aspects of GM food also need to be considered. Given that few developing countries have the technical ability to develop their own genetically modified food, use of GM food would mean dependence for food production upon multinational companies and international research agencies. This would undermine any food security policies and practices existing in many developing countries. The high costs of genetically modified seeds are also likely to squeeze many small and medium-sized farmers out of business. By analogy, the Green Revolution greatly increased food production, but more people are starving today despite world food yields consistently outstripping population growth since 1980. This is because hunger is not caused by inadequate levels of production but lack of access to land, money, and other resources. Introduction of GM foods without solving the underlying structural problems may only exacerbate the problems of hunger and food security.

Governments have taken different approaches to assess and manage the risks associated with the use of GM foods, with some of the most marked differences occurring between the United States and the European Union. In the United States, many GM products have been tested and commercially produced and marketed without any special labeling requirement. In the European Union, few products have been approved and a de facto moratorium has limited the production, import, and domestic sale of most GM crops. Only five EU countries grow GM crops at all—Spain, Portugal, the Czech Republic, Romania, and Slovakia—and in such tiny quantities that they accounted for less than 0.1% of global GM cultivation. Furthermore, the few GM foods that are allowed in the EU must be clearly labeled as such.

Why does Europe oppose GM food? There is no simple answer to this question, but Europe's fragmented politics, diverse landscapes, and smaller scale farming traditions have made it less compatible with the mass-farming techniques used in the Americas to produce GM crops. It costs the big agrochemical firms, such as Monsanto or Bayer, around 200 million dollars to develop the simplest GM seed. This investment gets recouped through aggressive marketing and monopoly ownership of seeds. Although a significant part of the EU opposition to GM foods is driven by economic forces, mainly a desire to protect local agriculture, the formal reasons given for banning GM foods are based on environmental and health cautionary concerns.

Genetic engineering has also been performed on animals, including fruit flies, mosquitoes, worms, sea anemones, fish, mice, cows, sheep, and primates (marmosets). Although most of the genetically modified animals were produced for research purposes, some were engineered to enhance production or food quality traits. One of the most interesting is genetically modified salmon for use in the aquaculture industry to increase the speed of development and potentially reduce fishing pressure on wild stocks. AquaBounty Technologies, a small biotechnology company with headquarters in Maynard, Massachusetts, United States, claims that their GM salmon can mature in half the time it takes non-GM salmon and achieves twice the size. Their hybrid Atlantic salmon incorporates a gene from a Chinook salmon. AquaBounty received FDA approval for the production, sale, and consumption of its genetically engineered salmon in November 2015. GM salmon is the first "transgenic" animal allowed into our food supply. It's also not labeled, so you might not even know you're eating it.

Dolly, born July 5, 1996, was a female domestic sheep, and the first mammal to be cloned from an adult cell. The cell used as the donor for the cloning of Dolly was taken from a mammary gland, and the production of a healthy clone therefore proved that a cell taken from a specific part of the body could recreate a whole individual. She was cloned by Ian Wilmut, Keith Campbell, and colleagues at the Roslin Institute, part of the University of Edinburgh, Scotland. She died from a progressive lung disease 5 months before her seventh birthday. She has been called "the world's most famous sheep."

In 2011, Chinese scientists generated 300 dairy cows genetically engineered with human genes to produce milk that would be the same as human breast milk. This could potentially benefit mothers who cannot produce breast milk but want their children to have breast milk rather than formula. Aside from milk production, the researchers claim these transgenic cows are identical to regular cows. In two additional breakthroughs, researchers have revealed that they have successfully created a calf whose milk could be drunk by people suffering from lactose intolerance and a second animal whose milk contains high levels of the healthy fat omega-3 fatty acids normally found in fish and nuts. People who are lactose intolerant lack the ability to digest milk properly and can cause digestion problems in sufferers. Omega-3 fatty acids are thought to be important for human health by helping to protect against heart disease and playing a role in brain function. These cows are part of a growing effort of scientists to make food and drink products from livestock healthier by genetically altering the animals. The plan is to create entire herds of these genetically modified cows, and have their milk hit markets by 2020.

Such research, however, is likely to inflame the debate about GM foods. Critics of the technology have reacted angrily to the research and questioned the safety of milk from genetically modified animals. Wendy Higgins, from the Humane Society International, said: "This simply isn't a morally responsible direction for farming to be heading in. Genetic modification of animals has an almost unique capacity to cause suffering and the welfare impacts on the animals produced can be both unpredictable and severe." Dr. Helen Wallace, director of Genewatch, added: "There is a question of food safety with GM livestock. As with all GM technology, there is a potential for unintended consequences as it is interfering with the natural biological production pathways of milk, so it could affect other nutrients or even have harmful effects."

Genetic engineering of humans is in its infancy. In February 2015, the British House of Commons and House of Lords approved the controversial technique of creating babies from three people, the mother, the father, and a female mitochondrial donor. This decision will enable the team at Newcastle Fertility Centre to apply for a license to offer the procedure to women at high risk of passing on inherited mitochondrial diseases. If they receive the license, the first baby with DNA from three different individuals could be born in 2017. The technique was developed by Doug Turnbull and his team at Newcastle University to help women with mitochondrial disease have healthy babies.

Mitochondria are the tiny compartments in the cytoplasm of nearly every cell of the body that convert food into useable energy. Mitochondria have their own DNA (mtDNA), which represents about 0.1% of the total DNA of the cell. Since the father's mtDNA is lost at fertilization, mtDNA is inherited exclusively from the mother. Inherited mtDNA diseases, which affect about 1 in 7500 children, can lead to brain damage, muscle wasting, heart failure, and blindness. The potential of a mother to transmit mtDNA disease to her child can be determined by DNA sequencing. If she has damaged mtDNA, the novel technology will give her the option of substituting her mtDNA with that of another woman.

Basically, the technique consists of collecting eggs from a potential mother with damaged mitochondria and a donor with healthy mitochondria. The nucleus which contains the vast majority of the genetic material is then removed from both eggs. The mother's nucleus is inserted into the donor egg, which can be fertilized by sperm from the father. It results in babies with 0.1% of their DNA from the second woman and is a permanent change that would be passed down through the generations.

Replacing damaged mitochondria is only the beginning of human genetic engineering. Techniques are now available to exchange or modify any gene on the human chromosome. There are clearly ethical implications that arise when modifying the human genome. Preventing genetic diseases by human genetic engineering is inevitable. But do we use it for cosmetic changes, such as eye color or for improving a desired athletic trait? An example is surgery, which we have performed for 100 years for disease purposes and is now widely used as a cosmetic tool. But there is an important difference. Changes in genes are transmitted to future generations. Genetic engineering of human embryos raises the fear of creating a eugenic driven human population. Decisions will have to be made, hopefully by informed and ethical people.

NOTES AND REFERENCES

82 Berg, P., Baltimore, D., Brenner, S., Roblin, R.O., Maxine, F., Singer, M.F., 1981. Summary statement of the Asilomar Conference on recombinant DNA molecules. Proc. Natl. Acad. Sci. USA 72, 1981–1984.

82 Berg, P., Singer, M.F., 1995. The recombinant DNA controversy: twenty years later. Proc. Natl. Acad. Sci. USA 92, 9011–9013.

83 NIH Guidelines for Research Involving Recombinant DNA Molecules. June 1976. Federal Register 41: 27911.

83 Kolata, G., 1994. Ethicists wary over new gene technique's consequences. The New York Times, p. C-10.

84 Ladisch, M.R., Kohlmann, K.L., 1992. Recombinant human insulin. Biotechnol. Prog. 8, 469–478.

84 Goeddel, D., et al., 1979. Expression in *Escherichia coli* of chemically synthesized genes for human insulin. Proc. Natl. Acad. Sci. USA 76, 106–110.

84 Baxter, L., et al., 2007. Recombinant growth hormone for children and adolescents with Turner syndrome. Cochrane Database Syst. Rev. 1, CD003887.

84 Pipe, S.W., 2008. Recombinant clotting factors. Thromb. Haemost. 99, 840–850.

85 Staff writers. Roche makes $43.7B bid for Genentech. Genetic Engineering & Biotechnology News. July 21, 2008.

85 Kevles, D.J., 1994. Ananda Chakrabarty wins a patent: biotechnology, law, and society. Hist. Stud. Phys. Biol. Sci. 25, 111–135.

85 Fraley, R.T., et al., 1983. Expression of bacterial genes in plant cells. Proc. Natl. Acad. Sci. USA 80, 4803–4807.

86 Shrawat, A., Lörz, H., 2006. Agrobacterium-mediated transformation of cereals: a promising approach crossing barriers. Plant Biotechnol. J. 4, 575–603.

86 Sanford, J.C., et al., 1987. Delivery of substances into cells and tissues using a particle bombardment process. Particulate Sci. Technol. 5, 27–37.

86 Lee, C.W., et al., 2009. *Agrobacterium tumefaciens* promotes tumor induction by modulating pathogen defense in *Arabidopsis thaliana*. Plant Cell 21, 2948–2962.

86 The active ingredient of the herbicide Roundup is Glyphosate (*N*-(phosphonomethyl) glycine), a broad-spectrum systemic herbicide used to kill weeds.

87 National Academy of Sciences, 2001. Transgenic Plants and World Agriculture. National Academy Press, Washington.

87 Sayre, R., et al., 2011. The BioCassava plus program: biofortification of cassava for sub-Saharan Africa. Ann. Rev. Plant Biol. 62, 251–272.

87 Ruiz-Lopez, N., et al., 2014. Successful high-level accumulation of fish oil omega-3 long-chain polyunsaturated fatty acids in a transgenic oilseed crop. Plant J. 77, 198–208.

88 Pennisi, E., 2013. The CRISPR craze. Science 341 (6148), 833–836.

88 Pollack, A., 2014. Seeking support, biotech food companies pledge transparency. The New York Times.

89 Losey, J., et al., 1999. Transgenic pollen harms monarch larvae. Nature 399, 214–215.

89 Jaenisch, R., Mintz, B., 1974. Simian virus 40 DNA sequences in DNA of healthy adult mice derived from preimplantation blastocysts injected with viral DNA. Proc. Natl. Acad. Sci. USA 71, 1250–1254.

90 Pollack, A., 2012. An entrepreneur bankrolls a genetically engineered salmon. The New York Times.

90 Wilmut, I., et al., 1997. Viable offspring derived from fetal and adult mammalian cells. Nature 385, 810–813.

91 Lopatto, E., 2012. Gene-modified cow makes hypoallergenic milk rich in protein. Bloomberg News.

Chapter 11

The Human Genome

We are here to celebrate the completion of the first survey of the entire human genome. Without a doubt, this is the most important, most wondrous map ever produced by human kind.

—Bill Clinton, June 26, 2000

Ten years after President Bill Clinton announced that the first draft of the human genome was completed, medicine has yet to see any large part of the promised benefits. For biologists, the genome has yielded one insightful surprise after another. But the primary goal of the $3 billion Human Genome Project — to ferret out the genetic roots of common diseases like cancer and Alzheimer's and then generate treatments — remains largely elusive. Indeed, after 10 years of effort, geneticists are almost back to square one in knowing where to look for the roots of common disease.

—Nicholas Wade, New York Times, June 12, 2010

The human genome is the entire genetic code of 3 billion letters required to create a human being. This genetic blueprint in each of our cells is encoded as DNA sequences within 23 chromosome pairs in the cell nuclei. If we are interested in the DNA blueprint for humans, or for any biological species, it is necessary to determine and analyze the precise sequence of bases (letters) in its DNA, using the methods described in Chapter 8. The biophysicist Robert Sinsheimer was one of the first to argue that the technology for determining the sequence of the human genome is available. In a letter to the President of the University of California, dated November 19, 1984, he wrote:

"It is an opportunity to play a major role in a historically unique event – the sequencing of the human genome. A genome is the complete set of DNA instructions for a human being. We know that the haploid human genome is composed of a sequence of some 3 billion base pairs (3×10^9).

A few months ago, I proposed to our biologists the question: could the human genome now be sequenced, with extant technique, and in a reasonable time (10 years)? If so, what scale of effort would be required? (Obviously, I had made a guess to the answer).

There answer is enclosed. It can be done. We need a building in which to house the Institute formed to carry out the project (approximately $25 million), and we

It's in Your DNA. http://dx.doi.org/10.1016/B978-0-12-812502-1.00011-1

need an operating budget of some $5 million/year (in current dollars). Not at all extraordinary.

Clearly, the human genome will be sequenced. It will be done, once and for all time, providing a permanent and priceless addition to our knowledge.

In addition to satisfying our scientific curiosity, this knowledge will provide deep insight into other questions of interest. It will have major medical implications: we know that literally thousands of human ailments have genetic bases, in whole or part.

This knowledge will also have major evolutionary implications. The biological differences between Homo sapiens and the chimpanzee are certainly due to the changes and rearrangements in the genomes of each as they have diverged from that of our common ancestor. To understand these changes will surely illuminate the ancient human quest to know what we are and where we came from."

Following Sinsheimer's advocacy for sequencing the human genome and several government sponsored workshops discussing the science and politics of the project, the United States Congress funded both the National Institutes of Health (NIH) and the Department of Energy (DOE) in 1988 to embark on further exploration of this idea. The two government agencies formalized an agreement by signing a Memorandum of Understanding to "coordinate research and technical activities related to the human genome."

One of the major advocates for sequencing the human genome was James Watson, who at the time was Director of the Woods Hole Laboratory. Interestingly, Watson's DNA was chosen to be the first human DNA to be sequenced, a fitting choice because of his major contribution to the double-helix structure of DNA.

In 1990, the initial planning stage was completed with the publication of a joint research plan, "Understanding Our Genetic Inheritance: *The Human Genome Project (HGP)*." The first 5 years, 1991–95, set out specific goals for what was then projected to be a 15-year research effort. The project was formally funded in 1990 at a cost of $3 billion. A large part of the initial effort was designed to improve and upscale the technology for accelerating the elucidation of the genome, building detailed genetic and physical maps, developing better, cheaper, and faster technologies for handling DNA, and mapping and sequencing the more modest-sized genomes of model organisms, all critical stepping stones on the path to initiating the large-scale sequencing of the human genome. In addition to the United States, the international consortium comprised scientists in the United Kingdom, France, Australia, Japan, and myriad other spontaneous relationships.

Competition from the private sector put much pressure on the HGP to work harder and faster to achieve its goals. However, there was a key difference between the goals of the private companies and the HGP. The sequence produced by the government-backed HGP mapped the entire, 3-billion-base human genome, in contrast to the private sector's emphasis on identifying specific genes. For this reason, when private sector competitors made announcements regarding human genome mapping, it was often in reference to mapping only certain regions of DNA with important genetic information. The key private competitor against the publicly funded Human Genome Project was Celera Genomics (*celer*: Latin for swift) led by Craig Venter, President and Chief Scientific Officer. Dr. Venter and his team of scientists have discovered and published about one half of all the human genes that have been sequenced.

The outcomes of the human genome projects have in them both good and bad news. The good news is that there has been amazing development of more rapid and less expensive DNA sequencing techniques that can be seen from the graph shown in Fig. 11.1.

As mentioned earlier, determination of the base sequence of the first human genome took 10 years (1990–2000) and cost of $3 billion. In June 2000, with great public fanfare, came the announcement that the human genome had in fact been sequenced, which was followed by the publication of the sequence of the genome's 3 billion base-pairs in the journal *Nature*. From 2001 to 2014, the cost per human genome decreased from $100 million to $6 thousand.

The earlier graph shows hypothetical data reflecting Moore's Law. Intel cofounder Gordon Moore's iconic observation that computing power tends to

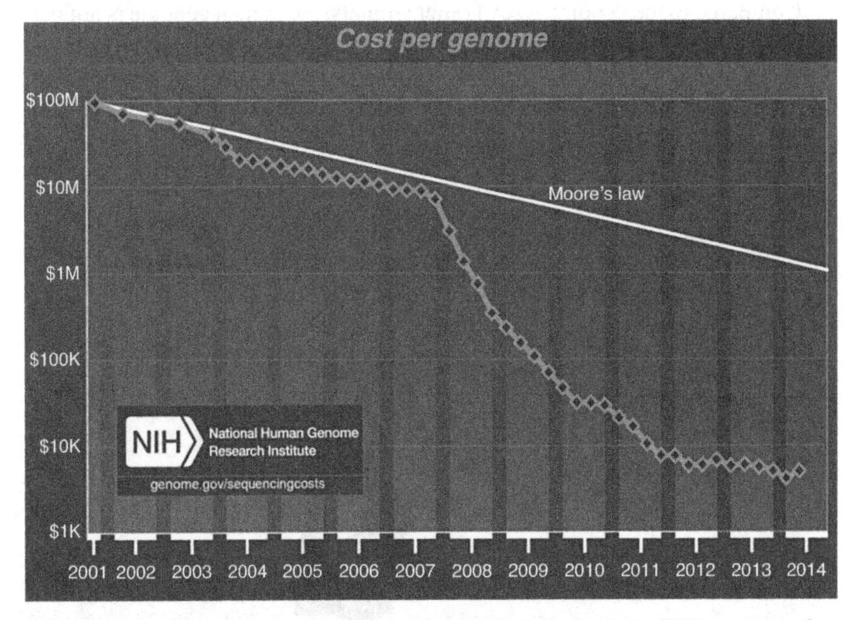

FIGURE 11.1 The graph shows how the cost of determining the entire DNA sequence of an individual human being has changed from 2001 to 2014.

double—and that its price therefore halves—every 2 years has held true for nearly 50 years with only minor revision. But as an exemplar of rapid change, it is the target of playful abuse from genome researchers. Over the past few years, scientists have compared the slope of Moore's law with the swiftly dropping costs of DNA sequencing. For a while they kept pace, but since about 2007, it has not even been close. The price of sequencing an average human genome has plummeted from about US$10 million to a few thousand dollars in just 6 years. That does not just outpace Moore's law—it makes the once-powerful predictor of unbridled progress look downright sedate. And just as the easy availability of personal computers changed the world, the breakneck pace of genome-technology development may soon revolutionize biomedical research.

As of 2014, about 228,000 human genomes have been sequenced. Scientists are optimistic that in the next 10 years, the price of sequencing a genome will drop to a few hundred dollars, and millions of babies will have their genomes sequenced at birth. It is anticipated that researchers will then have enough data to use in their analysis of how genetic variations manifest themselves in diseases and the era of personalized medicine will begin (Fig. 11.2).

More good news is that to date, sequencing human genomes has provided at least partial answers to some important biological question, such as: how does the human genome compare to other animals and to ancient humans, such as Neanderthals, and what is the variation in the DNA sequence of different modern humans?

Compared to the genome size of other animals, the human genome is not particularly large. The human genome contains 3 billion base pairs (bp). (Remember that the DNA is built of 2 strands, one opposite the other, creating pairs of complimentary nitrogen bases standing one opposite the other.) This size of

FIGURE 11.2 **The human genome containing 3 billion base pairs.**

3 billion bp fits within the range of other mammalian genomes (2–6 billion bp), so that genome size alone is not correlated with biological complexity. Surprisingly, some simpler animals have large genomes, for instance, the common toad *Buf-bufo* has a genome twice the size of the human genome and the even simpler single-celled organism *Amoeba proteus* has a genome size of 290 billion bp making it 100 times larger than the human genome.

One of the reasons for the lack of a relationship between genome size and biological complexity is that organisms can have multiple copies of the same gene, either by generating many copies of the same chromosomes (referred to as polyploidy) and/or by having many identical copies of DNA on the same chromosome. *Amoeba proteus,* for example, can have hundreds of copies of the same chromosomes, giving it a large DNA content without increasing the unique DNA information. One of the advantages conferred by having multiple copies of the same gene is the ability to diversify gene function over time. In other words, extra copies of genes might be used to produce mutations that generate new genes, leading to new opportunities in evolutionary selection.

How do humans genetically differ from one another? Most analyses of human genomes estimate that variations occur on the average of 1 out of 1000 base pairs. It follows that we are all, regardless of race, genetically 99.9% the same. However, no two humans are genetically identical; even monozygotic twins, who develop from one zygote (fertilized egg), have infrequent genetic differences due to mutations occurring during development and throughout life. These small differences in base pair sequences between individuals, even closely related individuals, are the key to techniques, such as genetic fingerprinting. DNA fingerprinting, also called DNA typing, is a technique used especially for identification (as for forensic purposes) by extracting and identifying the base-pair pattern in an individual's DNA.

To identify a specific individual, it is unnecessary to determine her/his entire genome. It is less expensive and more rapid to characterize individuals by their SNPs, (single nucleotide polymorphism and pronounced snips). What are SNPs? SNPs are DNA sequence variations occurring commonly within a population (e.g., 1%) in which a single base—A, T, C, or G—in the genome differs between members of the same biological species. For example, two sequenced DNA fragments from different individuals, AAGCCTA and AAGCTTA, contain a difference in a single nucleotide. By examining a number of SNPs, a particular individual can be identified with a probability of greater than 99.999%. SNPs have also been used to determine predisposition to the 30 leading medical causes of death and disability. In Chapter 16, I will discuss the role of SNP analyses in the early diagnosis of certain cancers and the potential therapeutic outcome in cancer patients.

Differences in the genome sequence of groups of humans have been used to study migration patterns of *Homo sapiens*. Based on fossil and DNA evidence, *H. sapiens* emerged in Africa approximately 200,000 years ago. The DNA evidence is based on the fact that Africans have more variation in their DNA than nonAfricans. We know that the older a racial group, the greater its DNA variability because of the longer the time it has to accumulate mutations. Humans only started to leave Africa between 60,000 and 70,000 years ago. Based on the available evidence, it is assumed that a small group of explorers (thus containing a limited DNA variability) ventured beyond Africa to colonize the Eurasian landmass likely across the Bab-el-Mandab Strait separating present-day Yemen from Djibouti. These early beachcombers expanded rapidly along the coast to India, and reached Southeast Asia and Australia by about 50,000 years ago.

Slightly later, a little after 60,000 years ago, a second group of Africans appears to have set out on an inland trek, leaving behind the certainties of life in the tropics to head out into the Middle East and southern Central Asia. From these base camps, they were poised to colonize the northern latitudes of Asia, Europe, and beyond.

When our ancestors began to migrate out of Africa around 60,000 years ago, they were not alone. At that time, at least two other species of hominid cousins walked the Eurasian landmass—Neanderthals and Denisovans. As our modern human ancestors migrated through Eurasia, they encountered the Neanderthals and interbred. Our ancestors were not a picky bunch. Overwhelming genetic evidence shows that *H. sapiens* had sex with Neanderthals, Denisovans, and other archaic relatives. Because of this, a small amount of Neanderthal DNA was introduced into the modern human gene pool. New findings indicate that Neanderthals became extinct between about 41,000 and 39,000 years ago.

Everyone living outside of Africa today has a small amount of Neanderthal in them, carried as a living relic of these ancient encounters. A team of scientists comparing the full genomes of the two species concluded that most Europeans and Asians have between 1% and 4% Neanderthal DNA. Indigenous sub-Saharan Africans have no Neanderthal DNA because their ancestors did not migrate through Eurasia.

Analysis of the DNA of Neanderthal man reveals that they contained two interesting mutations, one that gives rise to red hair and one that results in pale skin. These two mutations were not passed on to modern man. Red hair and fair complexions in modern man are the result of different mutations. At high latitudes where the Neanderthals resided in Europe, the UV radiation would have been minimal. And so fair skin, which has little protection from the sun's rays, would mean the individuals could absorb enough of the UV to produce vitamin D in sufficient amounts. Lighter skin is adaptive because vitamin D production depends on UV radiation. And like today's redheads, the researchers think Neanderthals featured a spectrum of red hair, from auburn to brilliant red to strawberry blonde (Fig. 11.3).

Around 20,000 years ago, during the last glacial period, a small group of Asian hunters headed north, entering the East Asian Arctic. At that time, vast ice sheets profoundly impacted Earth's climate, causing a drop in sea levels

FIGURE 11.3 Some Neanderthals may have had pale skin and red hair similar to that of some modern humans. *(Credit: Michael Hofreiter and Kurt Fiusterweier/MPG EVA.)*

by more than 300 ft. This exposed a land bridge that connected the Old World to the New World, joining Asia to the Americas. In crossing it, these nomadic hunters were following game herds from Siberia across what is, today the Bering Strait into Alaska. They had made the final great leap of the human journey. By 15,000 years ago they had penetrated the land south of the ice, and within another 1,000 years they had made it all the way to the tip of South America.

The story doesn't end there. The population explosion driven by the rise of agriculture around 10,000 years ago has left a dramatic impact on the human gene pool. The rise of empires, the astounding oceangoing voyages of the Polynesians, even the extraordinary increase in global migration over the past 500 years all left traces in our DNA. There are many human journey questions waiting to be asked and answered. What stories are waiting to be told in your own DNA?

Up to now we have discussed the good and interesting news coming out of the human genome project. Let us now turn to the unfulfilled hopes of the project. The main difficulty that we are confronting is that at present, 17 years after the sequence of the first human genome was announced, no major breakthrough in the prevention or treatment of disease has resulted from DNA sequence data. Part of the reason for this is that interpreting the meaning of DNA sequence data has been more complicated than anticipated. There are several reasons that can possibly explain this failure. To begin with there are more human proteins (250,000 to 1 million) than human protein-coding genes (approximately 22,000). Thus, the one gene–one enzyme hypothesis discussed in Chapter 6 is an oversimplification. One gene can encode more than one protein (even up to 1,000). How is this possible?

As was mentioned briefly at the end of Chapter 6, in humans and other eukaryotic organisms, transcription of the DNA to produce messenger RNA (mRNA) often involves splicing. mRNA splicing is a two-step process in which parts of the original RNA transcript are cut out and then the remaining parts are joined together to form mature mRNAs. By splicing together different fragments of the mRNA, different parts of a gene can be included within or excluded from the final mRNA. Consequently, the proteins translated from alternatively spliced mRNAs will contain differences in their amino acid sequence and, often, in their biological functions. Notably, alternative splicing allows the human genome to direct the synthesis of many more proteins than would be expected from its original protein-coding genes. In humans approximately 95% of the genes are alternatively spliced. A particular part of a gene may be included in an mRNA under some conditions or in a particular tissue, and omitted from the mRNA under different conditions or other tissues. The diversity of proteins is further increased by generating mRNAs by splicing together genetic information from different parts of different genes. Because of our ignorance of the rules governing splicing mRNA, DNA sequences do not currently allow us to predict with accuracy which proteins will actually be produced by a particular human gene.

Another problem in using genome data to discover ways to prevent or treat diseases is that not one gene or protein but multiple genes and proteins are responsible for most diseases, such as cancer, diabetes, obesity, cardiovascular disease, mental illness, and Alzheimer's disease. In these diseases, no single gene is responsible for more than a few percentage increases in the frequency of the disease. To prevent or treat these diseases it will be necessary to discover how these many different genes (sometimes hundreds) interact with each other and the environment to cause the disease.

Furthermore, the human genome contains a large amount of DNA that does not encode protein sequences. Some noncoding DNA is transcribed into functional noncoding RNA molecules (e.g., transfer RNA, ribosomal RNA, and regulatory RNAs), while others are not transcribed or give rise to RNA transcripts of unknown function. Initially, a large proportion of noncoding DNA that had no known biological function was referred to as "junk DNA," particularly in the lay press. However, we now know that this so-called junk DNA is not junk but contains the signals for the genome's "operating system"—the instructions for which are hidden in millions of locations around the genome.

It appears that the evolutionary rules for these regulatory DNA elements are different from those for protein-coding elements. Basically, the regulatory elements turn over a lot faster. So, whereas if you find a particular protein-coding gene in a human, you are going to find nearly the same gene in a mouse most of the time, and that rule just doesn't work for regulatory elements. It seems like evolution is driven more by changes in regulatory genes than genes that code for proteins.

Another unforeseen complexity arising out of the human genome project is the finding that in trying to use human genome data to understand disease and generate treatments is that a large amount of the genetic information of each one of us is not in the human genome. Each person contains thousands of different species of microbes and the combined numbers of their unique genes far exceed that of the human genome. These microbes contribute to human physiology and genetic variation. In the following Chapters 12 and 13, we will see how these beneficial microbial symbionts contribute to our health and disease.

At the onset of the Human Genome Project several ethical, legal, and social concerns were raised in regards to how increased knowledge of the human genome could be used to discriminate against people. One of the main concerns of most individuals was the fear that both employers and health insurance companies would refuse to hire individuals or refuse to provide insurance to people because of a health concern indicated by someone's genes. To address this concern, the United States passed the Health Insurance Portability and Accountability Act which protects against the unauthorized and nonconsensual release of individually identifiable health information to any entity not actively engaged in the provision of healthcare services to a patient.

Recently, the journal *Nature* published an article claiming that financial benefits derived from the Human Genome Project now hovers near US$1 trillion. "The economic impacts generated by the sequencing of the human genome are large, widespread and continue to grow," says Martin Grueber, the primary author of the report and a research leader at Battelle Memorial Institute. A decade after the project ended, the benefit now hovers near US$1 trillion. However, economists who were not involved in the study say that its numbers are not credible. Critics add that its basic approach is flawed because it quantifies the economic activity generated by the Human Genome Project instead of its impact on human health, which can be judged by metrics, such as patient outcomes and production of drugs and diagnostics. Robert Topel, an economist at the University of Chicago Booth School of Business in Illinois, says that the benefits of health research should not be measured by their effect on gross domestic product, productivity or jobs. "The question is: what health benefits have people got out of it, and what will they get in the future?" he says.

The scientist who probably best predicted the outcome of the human genome project was the Nobel laureate David Baltimore, who wrote: "We've got another century of work ahead of us to figure out how all these things fit together."

NOTES AND REFERENCES

95 Parfrey, L.W., Lahr, D.J., Katz, L.A., 2008. The dynamic nature of eukaryotic genomes. Mol. Biol. Evol. 4, 787–794.

96 http://www.genomenewsnetwork.org/articles/02_01/Sizing_genomes.shtml

96 http://www.library.georgetown.edu/DGViewe

96 Gregory, T.R., 2014. Animal Genome Size Database. Available from: http://www.genomesize.com

97 Wetterstrand, K.A., 1999. DNA sequencing costs: data from the NHGRI Genome Sequencing Program (GSP). Available from: www.genome.gov/sequencingcosts

98 Black, D.L., 2000. Protein diversity from alternative splicing: a challenge for bioinformatics and post-genome biology. Cell 103, 367–370.

98 Kidd, J.M., et al., 2008. Mapping and sequencing of structural variation from eight human genomes. Nature 453, 56–64.

99 Bruder, C.E.G., et al., 2008. Phenotypically concordant and discordant monozygotic twins display different DNA copy-number-variation profiles. Am. J. Hum. Genet. 82, 763–771.

99 Varela, M.A., Amos, W., 2010. Heterogeneous distribution of SNPs in the human genome: microsatellites as predictors of nucleotide diversity and divergence. Genomics 95, 151–159.

99 Adams, K.L., Wendel, J.F., 2005. Polyploidy and genome evolution in plants. Curr. Opin. Plant Biol. 8, 135–141.

100 Callaway, E., 2015. Neanderthals had outsize effect on human biology: from skin disorders to the immune system, sex with archaic species changed *Homo sapiens*. Nature 523, 512–513.

100 Culotta, E., 2007. Ancient DNA reveals Neandertals with red hair, fair complexions. Science 318, 546–547.

103 Goebel, T., Waters, M.R., O'Rourke, D.H., 2008. The late Pleistocene dispersal of modern humans in the Americas. Science 319, 1497–1502.

Chapter 12

Human Microbiome: We Are Not Alone

You could also ask who's in charge. Lots of people think, well, we're humans; we're the most intelligent and accomplished species; we're in charge. Bacteria may have a different outlook: more bacteria live and work in one linear centimeter of your lower colon than all the humans who have ever lived. That's what's going on in your digestive tract right now. Are we in charge, or are we simply hosts for bacteria? It all depends on your outlook.

—Neil deGrasse Tyson, an American astrophysicist, cosmologist and author

Mutually beneficial relationships between microbes and animals are a pervasive feature of life on our microbe-dominated planet. We are no exception: the total number of microbes that colonize our body surfaces exceeds our total number of somatic and germ cells by 10-fold, and the total number of microbial genes in our aggregate microbial communities is >100-fold greater than the number of genes in our human genome.

—Jeffrey Gordon, one of the pioneers of human microbiome research

We may think of ourselves as individuals, but each one of us is actually a group. All plants and animals, including humans, are complex symbiotic associations consisting of the host and abundant and diverse microorganisms. The sum of the associated microbes is referred to as the microbiota or microbiome. Often the number of bacteria in the microbiome and their combined genetic material exceed the number of host cells and host genes, respectively. Until recently, it was accepted that in humans there are 10 times more microbial cells than human cells (see earlier quote). However, recent facts suggest that the numbers of bacterial cells and hosts cells in humans are about the same. Because there are several thousand different species of bacteria associated with humans, the number of different bacterial genes in the human microbiome is >400-fold greater than the number of genes in our human genome. Before discussing how the microbiota affects the adaptation, development, health, behavior, and evolution of animals and plants, I would like to briefly describe the historical developments in microbiology which made possible our current understanding of organisms as complex systems.

It's in Your DNA. http://dx.doi.org/10.1016/B978-0-12-812502-1.00012-3

Microbiology, like most scientific disciplines, has advanced by observations, asking interesting questions and developing techniques to understand these observations and answer these questions. Between 1653 and 1673, Antony van Leeuwenhoek developed the curious hobby of constructing microscopes. Although he was not the first to build a microscope, his microscopes were the finest of that time. Leeuwenhoek patiently improved his microscopes and techniques of observation for 20 years before he reported his results in the form of letters, written in "Nether-Dutch" (Leeuwenhoek knew no Latin or English), to the secretary of the Royal Society of England. After translation, the letters were published in the Philosophical Transactions of the Royal Society. His 39th letter, dated September 17, 1683, which describes bacteria in the human mouth, is an excellent illustration of his charming style and accurate observational ability:

> *Tis my wont of a morning to rub my teeth with salt, and then swill my mouth out with water: and often, after eating, to clean my back teeth with a toothpick, as well as rubbing them hard with a cloth: therefore my teeth, back and front, remain as clean and white as falleth to the lot of few men of my years, and my gums (no matter how hard the salt be that I rub them with) never start bleeding. Yet notwithstanding, my teeth are not so cleaned thereby, but what there sticketh or groweth between some of my front ones and my grinders (whenever I inspected them with a magnifying mirror), a little white matter, which is as thick as if 'twere batter. I have therefore mixed it, divers times, with clean rainwater, and also with spittle, that I took out of my mouth, after ridding it of air bubbles (lest the bubbles should make any motion in the spittle): and I then most always saw, with great wonder, that in the said matter there were many very little animalcules, very prettily a-moving. The biggest sort shot through the water (or spittle) like a pike does through water. The second sort oft-times spun round like a top, and every now and then took off in a directed course...........*

Following van Leeuwenhoek's initial observations of microbes, it took 200 years for microbiology to develop into an experimental science. One of the main reasons for this long delay was there was no way to isolate bacteria. Let us see how this major difficulty was overcome. The liquid suspensions that Leeuwenhoek and other pioneers in microbiology examined under the microscope contained a wide assortment of microbes of varying sizes and shapes. Was this because a single microbe could exist in various forms, or was it the result of a mixture of different organisms, each having a single form? To answer this question and to test the theory that a particular microbe is responsible for a particular disease (germ theory of disease), it became necessary to develop a way of obtaining pure cultures of microbes. A pure culture is a culture grown from a single kind of microorganism and is free from other organisms. It can be a culture in a liquid form grown in a flask or it can be a culture grown in a so-called solid form—on a plate with hard jellified agar.

In 1881, the German bacteriologist Robert Koch developed the simple and efficient "streak-on-agar medium" technique for obtaining pure cultures that is still used today. Basically, the procedure involves dipping a sterilized metal or plastic loop into a mixed culture of bacteria and then streaking the bacteria over a solid agar surface containing nutrients. Toward the end of the streak, single, well-separated microbes are deposited. When incubated, the single cell divides every few hours to give 2, 4, 8, 16, etc. cells, eventually forming a visible colony of identical microbes (a pure culture) in a few days. The technique proved so valuable that by 1900, 21 microbes that caused diseases were obtained in pure culture and identified.

At about the same time that Robert Koch, Louis Pasteur and others were employing pure cultures to isolate and characterize pathogenic bacteria, a second school of microbiology, led by the Russian microbiologist and ecologist Sergei Winogradsky and the Dutch microbiologist Martinus Beijerinck, argued that the prime focus of microbiology should be the role of microbes in the turnover of chemical matter on the planet. The two schools, Koch/Pasteur as opposed to Winogradsky/Beijerinck, stood for different viewpoints. The former concentrated on diseases and pure cultures, while the latter focused on the environment and the interaction within mixtures of microorganisms that live on different chemical substances. The conflict between the two schools of microbiology was won at the time by the Koch/Pasteur school mainly because of its power in combatting infectious disease. The students of the Koch/Pasteur school received the academic appointments, became the teachers of microbiology, got most of the research funds, and became the editors of the journals. As we will discuss further in this chapter, it was not until the end of the 20th century that the environmental school began to find its rightful place in microbiology.

During most of the 20th century, microbiological research continued to focus on isolating microbes that were causative agents of disease and studying their transmission and mechanism of pathogenesis. Subsequently and not less important, these pure culture techniques were applied as model microbial systems to discover the underlining mechanisms of biochemistry and genetics. To achieve these goals, essentially all research in microbiology was performed with a limited number of pure bacterial cultures under defined laboratory conditions. Insufficient tools were available at the time to study mixtures of microorganisms and their ecological and evolutionary relationships.

The recent advent of techniques for sequencing DNA described in Chapter 8, coupled with the ability to use computers for analyzing the complex data in addition to the cumulative knowledge that had been gained on pure bacterial cultures opened the door for studying and understanding the vast microbial richness existing in our world. This has led to a merger of microbiology with botany, zoology, human physiology, and medicine, with the microorganisms at the base of global bioecosystems. Moreover, it becomes clear today that comprehending the vast role of microorganisms in nature will change the way we understand biology including the principles of ecology, physiology, embryology, immunity, and evolutionary biology. Microbes are not only the causative agents of some diseases or tools for studying the principles of biochemistry and genetics, but they are a fundamental part of all animals and plants, as well as the major organisms that are present in natural soils and water. To understand the role of microbes in humans, animals, and plants we have to first discuss the subject of symbiosis.

Biologists have studied symbioses, "the living together of different species," since the end of the 19th century. One of the first symbiotic organisms to be discovered was the lichen, which is composed of a photosynthetic algae and a fungus. The fungus benefits from the symbiotic relation because algae produce food by photosynthesis (the conversion of light energy into useful chemical energy and use of the chemical energy to form sugars and oxygen from carbon dioxide and water). The algae benefit by being protected from the environment by the fungus, which also gathers moisture and nutrients from the surroundings, and provide an anchor for it.

The symbiotic system is usually constructed from a large partner termed the host and smaller partners called symbionts. In the biological world, symbiosis can occur between varieties of different organisms, for example, between the clownfish and sea anemone, where clownfish that dwell among the tentacles of sea anemone provide nutrients to the sea anemone, and the sea anemone's stinging cells protect the clownfish from predators. Another example is the cleaning symbiosis between birds and mammals, where the bird (the cleaner) removes and eats parasites from the surface of the mammal (the client).

Until recently most of the important research on symbiosis involving microorganisms has been carried out with a small number of model symbiotic systems involving a host and a single symbiont. These model systems include nitrogen-fixing bacteria (*Rhizobia*) that live in root nodules of legume plants, single-celled algae (commonly referred to as zooxanthallae) inside reef-building corals, bacteria (*Buchnera*) that produce essential amino acids for aphids, and marine bacteria (*Vibrio fischeri*) that produces light for squids. Studies on these systems and others have led to a greater understanding of how symbioses are established, maintained and how the two partners coevolved.

Among the microbial symbionts one finds endosymbionts and exosymbionts, referring to microorganisms living inside or outside host cells, respectively. Symbioses can take many forms; most of them are such that both host and symbiont benefit from the interaction and are defined as different levels of mutualism. A different kind of symbiosis, when the symbionts benefit and the hosts suffer damage, is termed parasitism (or pathogenesis). Another type of symbiosis is commensalism which is accepted to be a close association between two or more dissimilar organisms where the association is advantageous to one and doesn't affect the other(s). In fact, the type of symbiosis is often context dependent. *Escherichia coli* is considered a commensal symbiont in the human gut, but if it moves to the bladder or urinary tract it can become a pathogen, causing severe infections.

During the last 20 years, there has been a dramatic change in how symbioses are viewed in biology. Rather than being restricted to a few interesting cases, it is now accepted that symbiosis is a general property of the biological world.

All plants and animals, including humans, are complex organisms, each one containing hundreds or thousands of different microbial species, including bacteria, fungi, and viruses. There are no natural germ-free animals or plants. In the remainder of this chapter, I will discuss the abundance and diversity of symbiotic microorganisms in humans, animals, and plants. Chapter 13 discusses how these symbionts affect the health and fitness of their hosts. In Chapter 15, the previously underappreciated key roles of microbiotas in the evolution of animals and plants will be presented.

Discovery of the large numbers and types of microbes associated with animals and plants depended on the development of DNA technology for analyzing them without having to isolate and grow them. Today, a scientist simply isolates the DNA from an organism, carries out a PCR reaction to amplify all of the genes that code for bacterial ribosomal RNA and then sequence these genes. (Sequencing is often performed by dedicated companies that perform the service for a fee.) Using available data bases of DNA sequences for the different microbial ribosomal genes, each PCR reaction sequence can be assigned to a specific microbial group. Table 12.1 provides some examples of the number of different bacterial species associated with animals (invertebrates and vertebrates) and plants.

Sponges range in size from a few millimeters to more than a meter in diameter. It is interesting that these organisms, which may be the most ancient animal that still exists, contain the highest concentration of microbial symbionts. Bacterial symbionts can comprise as much as 40% of sponge tissue volume, with concentrations in excess of 1 billion microbial cells per millilter of sponge tissue. Sponges contain approximately 3000 different bacterial species. High numbers of bacterial species are also found in other invertebrates, such as corals, hydra, and insects.

Symbioses between diverse microbiota and vertebrates have been studied in a variety of animals, including humans, apes, ruminants, rats, fowl, fish, and snakes. Because of its importance to human health, bacteria associated with the human gut have been the most intensely studied microbial ecosystem. It has been estimated that the microbes in a typical human healthy gut comprise 100 trillion (10^{14}) cells, approximately the same number as human cells in the body. Most of these microbes are not merely in transit but rather inhabit defined niches in the gut.

Determinations based on DNA techniques claim a minimum number of 5,000 and possibly upward of 20,000 different bacterial species in the human gut. These diverse bacteria encode 400-fold more unique genes than the human genome. It is rather amazing to realize that our bacteria possess more than 99% of the genetic information within each one of us. There is considerable variation in the types of bacteria in the gut of different individuals. Each person possesses their own personalized fingerprint of gut microbiota. Even identical twins have different microbiota. Human microbial communities are affected by factors, such as lifestyle, dietary patterns, antibiotic usage, and gender. Furthermore, the human intestinal microbiota undergoes maturation from birth to adulthood and is further altered with aging.

In addition to the gastrointestinal tract, there is a high abundance and diversity of microbes on all surfaces of the human body, including the oral cavity,

TABLE 12.1 Numbers of Microbial Species Associated With Different Animals and Plants

Host	Estimated number of bacterial species
Invertebrate animals	
Marine sponge	2,996
Coral	2,050
Termite gut	800
Honey bee	700
Hydra	350
Fruit fly	209
Vertebrate animals	
Human gut	5,700
Human mouth	1,500
Human skin	2,000
Great ape gut	8,914
Cow rumen	5,271
Zebrafish	178
Burmese python snake	500
House sparrow	4,436
Chicken	243
Plants	
Phylosphere: stems, leaves, flowers, fruits	252
Rhizosphere: root system	30,000
Marine green alga	1,061
Carnivorous pitcher plant	1,000

skin, nasal cavity, pharynx, esophagus, vagina, and urinary tract. The total number of bacteria on the surface of an average human (approximately 2 m^2) has been estimated to be 2 trillion (2×10^{12}), corresponding to 10^8 per cm^2. Most are found in the outer layers of the skin (epidermis) and the upper parts of hair follicles. The skin provides three very different ecological areas for the microbes, moist, dry, and sebaceous, each with their distinct microbial community. DNA analyses of skin samples have revealed about 1000 different bacterial species. Interestingly, a comparison of the skin microbiota of South American Amerindians and residents of the United States indicated that ethnicity, lifestyle, and geography can influence the structure of human skin bacterial communities (Fig. 12.1).

FIGURE 12.1 There is 400 times more genetic information in the human microbiome than in the human genome.

In an interesting comparative study of the microbiotas of the colon of primates (including *Homo sapiens*), it was found that the composition of these microbial communities is completely congruent with the known evolutionary relationships of the hosts, that is, the host primates and their microbiota coevolved. Although the gut is continuously seeded by bacteria from external sources and influenced by diet and other factors, the study established that over evolutionary timescales, the composition of the gut microbiota among great ape species has diverged in a manner consistent with the transmission of genes from parent to offspring (vertical inheritance).

Plants are an essential part of life on this planet, and symbiotic microorganisms play a central role in their fitness. Microorganisms supply plants with nutrients, play a role in development of their root systems, and protect them against pathogens and other environmental stress conditions. Microbes are found in three main locations in plants: the phyllosphere, namely the upper parts of the plant, leafs, stems, flowers, and fruits, the rhizosphere, the region of soil in the vicinity of plant roots, and inside plant cells (endophytes).

Most of the higher plant species enter into a mutualistic root symbiosis with fungi. Development of the fungi begins with invasion of the plant root by fungi in the soil; growth of the fungus toward the root is stimulated by nutrients released by the plant into the soil. The fungi penetrate the root cells and multiply inside the plant cells. In addition to fungi, many bacterial species interact with plant roots. The rhizosphere, which is the narrow zone of soil that is influenced by root secretions, can contain up to a hundred billion (10^{10}) microbial cells per gram root and contain more than 30,000 different microbial species. The collective genome of this microbial community is much larger than that of the plant. Rhizosphere microbial communities are different in different types of plants, at different developmental stages of a given plant, and also different from microorganisms present in surrounding soil.

An interesting example of the role of microbiota in plants is the American carnivorous pitcher plant (*Sarracenia*). The ability of these plants to digest insect prey is facilitated by microbial associations. Each pitcher (a modified leaf) of the plant contains a microcosm composed of different microorganisms, which together with plant enzymes kill and digest the insect prey, providing nutrients for the plant. Pitcher plants contain c. 1000 different bacterial species.

How do animals acquire their symbiotic microorganisms? Most importantly, offspring receive, to begin with, a diverse set of microbes from their parents. This initial input is probably the most important imprint of diversity in an offspring. Later on, this first basic microbiota can be supplemented with microorganisms from the environment throughout life.

In humans, like most mammals, most of the microbes are obtained from maternal vaginal and fecal bacteria as the newborn passes through the birth canal. Babies delivered vaginally are covered in a film of microbes. Included in the mix are bacteria that help babies digest their first meal. Babies delivered by cesarean section are colonized mainly by skin microbes from the hospital environment. These differences persist throughout infancy and seem to have an effect on the child's health. For example, infants delivered by caesarean section have a higher risk of asthma and twofold higher odds of childhood obesity than vaginally delivered infants.

Although traditionally thought to be sterile, recent studies have shown that human breast milk contains hundreds of different beneficial bacterial species, including *Bifidobacterium*, staphylococci, streptococci, and lactic acid bacteria. These bacteria come from inside the breast as an integral component of the milk. Not only does human milk provide bacteria for the infant gut, but remarkably it provides food specific for these microbes. Human mothers' milk contains complex sugars that are not digestible by the infant. Rather, they serve as food for certain bacterial symbionts that help the infants' bodies develop.

The genome of one of the most common bacteria found in human milk, *Bifidobacterium longum*, has been sequenced and shown to contain a remarkable region of DNA—a series of genes linked together and dedicated to the intake and digestion of complex sugars found specifically in the earliest secretions of mother's milk. These data indicate a remarkable coevolution between this symbiotic bacterium and its host, suggesting that the host product (human milk) and the microbial genome enabling the bacteria to use this product have reciprocally formed each other. Thus, there was coevolution of human and microbes for the purpose of gut colonization by the microbes, a colonization that would benefit both.

Many young animals, including iguanas, rabbits, horses, elephants, pandas, koalas, and hippos eat the feces of their mothers (a process called coprophagy), thereby obtaining the bacteria required to properly digest vegetation found in their environment. When they are born, their nutrition is based only on mother's milk. During this time their intestines do not contain the bacteria needed to digest plant food and without them they would be unable to extract sufficient nutritional value from plants. Koalas use a special adaptation of coprophagy; development of the young, referred to as Joeys, in the pouch is very slow with the Joey remaining in the pouch for 5–6 months and relying only on its mother's milk. When the Joey is approximately 5 months of age, the mother produces a second type of feces (known as pap) which the Joey eats over several days up to a week. This facilitates introduction of the appropriate gut microbiota into the developing juvenile's digestive system, and allows for the digestion of the eucalyptus leaves, the koala's food source, and the eventual weaning from the mother's milk.

What is the basis for the high diversity of the microbiota of animals and plants? Why does the human body, for example, contain thousands of different species of bacteria and other microbes? Why don't a few types out compete the others and drive them to extinction in the human habitat? Let us consider some forces that would maintain a high diversity. The main reason for that is that many microorganisms are specialists. Given that hosts provide a variety of different niches that can change with the developmental stage of the host, the diet, temperature, and other environmental factors, a diverse microbial community is established, with different microbial strains filling the different niches. This microbial diversity, and therefore its versatility, may allow the holobiont as a whole to function more optimally and adapt more rapidly to changing conditions.

The idea that microbial diversity can play a critical role under conditions of fluctuating environments has been referred to as the "insurance policy hypothesis." There are two parts to this hypothesis: the first is that a single animal or plant can contain in its microbiota a reservoir of rare microbes which have the potential to increase in number when conditions change and are favorable for them and in that way assist that specific organism to adapt and survive in the new environment. The second part is that the presence of rare microbes in a population of organisms of the same host species, but not necessarily in all members of the species, can help insure the long-term survival of the host species. In essence, the "insurance policy hypothesis" offers another layer to biodiversity.

Another factor that may contribute to bacterial diversity is bacterial viruses (phages). High concentrations of phages are present in animal and plant tissues. The human gut, for example, contains 1200 different phages. If any bacterial strain becomes abundant, it has a high probability of being attacked by a phage because the collision frequency of a phage and its bacterial host is directly proportional to the host concentration. The higher the concentration of a particular bacterium, the greater the chance it will be killed by a phage. This concept is referred to as the "kill the winners" hypothesis. Since this mechanism removes abundant bacterial species, it selects for rare types and thus maintains high diversity of microbes.

The next chapter will discuss how the abundant and diverse microbiotas play a key role in the health and fitness of humans, animals, and plants.

NOTES AND REFERENCES

105 Metchnikoff, E., 1907. In: Mitchell, P.C. (Ed.), The Prolongation of Life. Optimistic Studies. Heinemann, London.

105 The term microbiome was coined by Lederberg and McCray (Scientist 15, 8 2001) to describe the sum of the genetic information of the microbiota.

107 Dworkin, M., 2012. Sergei Winogradsky: a founder of modern microbiology and the first microbial ecologist. FEMS Microbiol. Rev. 36, 364–379.

108 Ott, T., et al., 2009. Absence of symbiotic leghemoglobins alters bacteroid and plant cell differentiation during development of Lotus japonicus root nodules. Mol. Plant Microbe. Interact. 22, 800–808.

108 Buddemeier, R.W., et al., 2004. The adaptive hypothesis of bleaching. In: Rosenberg, E., Loya, Y. (Eds.), Coral Health and Disease. Springer-Verlag, New York, NY, pp. 427–444.

108 Wilson, A.C.C., et al., 2010. Genomic insight into the amino acid relations of the pea aphid *Acyrthosiphon pisum* with its symbiotic bacterium *Buchnera aphidicola*. Insect Mol. Biol. 19, 249–258.

109 Schmitt, S., et al., 2012. Assessing the complex sponge microbiota: core, variable and species-specific bacterial communities in marine sponges. ISME J. 6, 564–576.

109 Frank, D.N., Pace, N.R., 2008. Gastrointestinal microbiology enters the metagenomics era. Curr. Opin. Gastroenterol. 24, 4–10.

109 Qin, J., et al., 2010. A human gut microbial gene catalogue established by metagenomic sequencing. Nature 464, 59–65.

109 Faith, J.J., et al., 2013. The long-term stability of the human gut microbiota. Science 341, 1237439.

110 Markle, J.G.M., et al., 2013. Sex differences in the gut microbiome drive hormone-dependent regulation of autoimmunity. Science 339, 1084–1088.

110 Zarco, M.F., et al., 2012. The oral microbiome in health and disease and the potential impact on personalized dental medicine. Oral Dis. 18, 109–120.

110 Blaser, M.J., et al., 2013. Distinct cutaneous bacterial assemblages in a sampling of South American Amerindians and US residents. ISME J. 7, 85–95.

110 Grice, E.A., et al., 2009. Topographical and temporal diversity of the human skin microbiome. Science 324, 1190–1192.

111 Ochman, H., et al., 2010. Evolutionary relationships of wild hominids recapitulated by gut microbial communities. PLoS Biol. 8, e1000546.

111 Bulgarelli, D., et al., 2013. Structure and functions of the bacterial microbiota of plants. Annu. Rev. Plant Biol. 64, 807–838.

112 Dominguez-Bello, M.G., et al., 2010. Delivery mode shapes the acquisition and structure of the initial microbiota across multiple body habitats in newborns. Proc. Natl. Acad. Sci. USA 107, 11971–11975.

113 Yachi, S., Loreau, M., 1999. Biodiversity and ecosystem productivity in a fluctuating environment: the insurance hypothesis. Proc. Natl. Acad. Sci. USA 96, 1463–1468.

113 Breitbart, M., et al., 2003. Metagenomic analyses of an uncultured viral community from human feces. J. Bacteriol. 185, 6220–6223.

113 Thingstad, T.F., Lignell, R., 1997. Theoretical models for the control of bacterial growth rate, abundance, diversity and carbon demand. Aquat. Microb. Ecol. 13, 19–27.

113 Chiu, L., Gilbert, S.F., 2015. The birth of the holobiont: multi-species birthing through mutual scaffolding and niche construction. Biosemiotics 8, 191–210.

Chapter 13

Contribution of Microbes to the Health of Humans, Animals, and Plants

A reader who has little knowledge of such matters may be surprised by my recommendation to absorb large quantities of microbes, as a general belief is that microbes are harmful. This belief is erroneous. There are many useful microbes, amongst which the lactic bacilli have an honorable place.

—Elie Metchnikoff, 1907

Pathogenicity is not the rule. Indeed, it occurs so infrequently and involves such a relatively small number of species, considering the huge population of bacteria on earth, that it has a freakish aspect. Disease usually results from an inconclusive negotiation for symbiosis, an overstepping of the line by one side or the other, a biological misinterpretation of borders.

—Lewis Thomas, 1974

As documented in the previous chapter, all animals and plants contain large numbers and types of symbiotic microorganisms, including bacteria, yeasts, and viruses. We now address the question: how do these resident microbes contribute to the adaptation, behavior, development, and overall health of their hosts? Because of the importance of beneficial microbes, the title of this book, "It's in Your DNA," refers not only to the genetic information in the host chromosomes but also in the enormous diversity of DNA information present in your resident symbionts.

The concept of fitness is central to the Darwinian theory of evolution by natural selection—the fittest survive and transmit their advantageous traits to their offspring, but there are many definitions of biological fitness. The following qualitative definition of "Darwinian fitness" will suffice for the purposes of this chapter: the ability of an organism to survive and reproduce in a specified environment and population. The phrase "in a specified environment and population" should be emphasized because there is no absolute fitness, only relative fitness within a particular environment and population. For example, certain remote islands in the Pacific Ocean contain a high proportion of wingless insects, presumably because periodic high winds blow the winged insects

It's in Your DNA. http://dx.doi.org/10.1016/B978-0-12-812502-1.00013-5

FIGURE 13.1 **Your symbiotic microbes protect you against pathogens, provide you with nutrients and influence your development and behavior.**

out to sea. Thus, wings are a positive fitness trait for insects on the continents from where the insects migrated, but a negative fitness trait on the islands.

Symbioses between microbes and their hosts benefit all partners. The microbes are provided a comfortable home, containing abundant nutrients and protection against predators. Although the idea that resident microbial communities are important contributors to the well-being of their hosts is not new, we are now developing a much broader and deeper appreciation of the wide range of functions that are performed by symbiotic microorganisms. Although the term "outsourcing" has been applied to describe functions that microbiotas contribute to their hosts, the term "insourcing" is more appropriate because microbiotas are an integral part of organisms. If we want to summarize the general areas in which the symbiotic microbes contribute to their hosts, it would be as follows: protection against pathogens and toxic materials, providing nutrients to hosts, and influencing important physiological systems, such as developmental processes, behavior, obesity, and mate selection. It should be emphasized that the presently known contributions of microbiotas to their hosts are probably only the tip of the iceberg because every month new and exciting articles appear in the scientific literature and popular press showing unexpected roles symbionts play in the health of their hosts (Fig. 13.1).

Symbiotic microbes protect against pathogens. Protection against pathogens is one of the most general and important contributions of the resident microbiota to the health of humans, animals, and plants. The initial demonstration that symbiotic microorganisms play a key role in preventing infection by disease-causing microbes involved the use of germ-free animal models. Germ-free animals are born in aseptic conditions, which may include removal from the mother by cesarean section and immediate transfer of the newborn to an isolator where all incoming air, food, and water are sterilized. Such animals are normally reared in a sterile or microbial-controlled laboratory environment, and they are only exposed to those microorganisms that the researchers wish to have present in the animal. Germ-free animals are a 1000 times more sensitive to infection and death following oral administration of a pathogen than conventional animals (those containing the normal microbiota). These experiments include infection of guinea pigs with the bacterial pathogen *Shigella flexneri*, mice with pathogenic *Vibrio cholera* or the virus Influenza A, and rabbits with the pathogen *Bacteroides vulgatus*.

Further evidence that resident microbes protect animals against infection came from an experiment in which germ-free mice were treated with *Bifidobacterium longum*, part of the normal microbiota, and then infected with the pathogen *Salmonella typhimurium*. The mice that received *Bifidobacterium* survived, whereas the control group (treated only with *Salmonella*) all died within a few days. Similar results have been obtained with other germ-free animals and plants.

In humans, the normal microbiota has been shown to protect against infection by pathogens in the oral cavity, the intestine, the skin, and the vaginal epithelium. The fact that there is an increased frequency of infection by pathogenic yeast following antibiotic therapy is consistent with this concept. Removal of the resident bacteria by antibiotics allows pathogenic yeast of the genus *Candida* to cause oral and vaginal infections known as candidiasis. In the case of the vaginal epithelium, it is known that the resident bacteria produce lactic acid which inhibits the growth of yeast. One of the strongest arguments that beneficial bacteria prevent disease in humans comes from recent probiotic experiments. Probiotics are live microorganisms which, when administered in adequate amounts, confer a health benefit on the host. Several trials indicate that probiotics are effective in prevention and treatment of gastroenteritis.

With the alarming increase in antibiotic-resistant bacterial pathogens, there is an increasing interest in applying beneficial microbes (probiotics) to treat serious infections. In this regard, probiotics have been effective in treating recurrent diarrhea caused by the pathogen *Clostridium difficile*. In human trials, patients who had not responded to antibiotic treatment were successfully treated with a nasogastrically administered (a procedure during which a thin, plastic tube is inserted through the nostril, down the esophagus, and into the stomach) liquefied blend of feces obtained from healthy related donors. In a patient who was infused with her husband's microbiota, there was a rapid and prolonged change in the bacterial composition of her gut microbiota, with *Clostridium* disappearing and defecation frequency returning to normal. In a recent review article summing up results from more than 300 patients, it was concluded that fecal transplants can cure 92% of people with recurring *C. difficile* infections for which antibiotics proved ineffective. It may seem obnoxious to you to be treated with someone else's feces, but fecal transplantation is lifesaving for patients who are critically ill with *C. difficile* infection. In 2016, a new pill was created by a team of MIT-trained researchers, providing fecal transplant therapy without requiring the traditional, more invasive stool delivery methods.

The newborn infant not only tolerates but requires colonization by symbiotic microbes for development and health. Human milk provides not only all the nutrients needed to satisfy the growth requirements for babies, but also contains specialized complex sugars that stimulate the growth of select beneficial bacteria. This three-way relationship between mother, child, and gut microbes protects the infant against infection by hostile pathogens at a time when the infant has an untrained immune system and produces low amounts of caustic stomach acid, which in adults kills most pathogenic bacteria.

Symbiotic microbes detoxify toxicants. Animals, including man, are exposed to many environmental toxic materials that are known to cause disease and death. Several microorganisms present in the digestive tract of animals have the ability to bind and detoxify some of these substances. For example, plant and fungal toxins are commonly found on corn and can be detoxified by microorganisms in the rumen of cows and sheep. The highly toxic chemical hydrazine (used in agricultural chemicals, photography, boiler water treatment for corrosion protection, and textile dyes) is broken down by bacteria in the human gut. Also, toxic metals can be removed by microbiota. Studies in mice showed that gut bacteria provide the first line of defense against the toxic metals chromium, lead, and cadmium. Two common probiotics, *Lactobacillus rhamnosus* and *Propionibacterium freudenreichii*, have been shown to bind these metals and remove them from the colon.

Another good example of detoxification by microbes in the human colon is the breakdown of oxalic acid by the bacterium *Oxalobacter formigenes*. Oxalic acid is present in a wide range of animal feeds and human foods and beverages, including almonds, cashews, buckwheat, tea, coffee, chocolate, soy protein, beets, chocolate, rhubarb, spinach, and other fruits and vegetables. Young and fresh vegetables, such as baby spinach, are less likely to have oxalic acid. The ingestion of oxalic acid can result in precipitation of calcium oxalate in the kidneys (kidney stone disease). The ability of *Oxalobacter* to degrade dietary oxalic acid prompted its successful use in clinical trials as a therapeutic and prophylactic option in treating kidney stones and associated renal failure. In general, gut microbiota are helpful in preventing adverse outcomes following inadvertent environmental exposure to toxic compounds.

Symbiotic microbes provide nutrients to their hosts. Contribution of nutrition to their hosts is another general property of microbiota. In humans, the gut microbiota is a complex ecosystem that plays an essential role in the breakdown of dietary fiber, and production of vitamins and amino acids. Certain bacteria in the gut breakdown dietary fiber, that part of plant polysaccharides in our diet that is not metabolized in the upper digestive tract because the human genome does not encode adequate enzymes. In general, many herbivores and omnivores would not be able to extract energy from plant fiber if it were not for their gut microbiota. During times of food shortage, symbioses with fiber-degrading bacteria provide a clear fitness advantage. For a long time, it has been known that a high fiber diet leads to a reduction in blood pressure, blood cholesterol, and cardiovascular diseases. One of the factors that play a role in achieving this effect is short-chain fatty acids (SCFAs), produced by gut bacteria by fermentation of fiber and a form of starch (resistant starch) that escapes from digestion in the small intestine.

The principal SCFAs produced via bacterial fermentation are acetate, propionate, and butyrate, present in the human colon in the approximate molar ratio of 60:20:20. SCFAs show important and pleiotropic functional effects via their influence on key regulatory proteins. SCFAs affect cell proliferation and function and microbe-to-host relationship, have antiinflammatory, antitumorigenic properties, and are major players in energy metabolism and maintenance and function of the immune system. Given the many effects of SCFAs, they also play a role in affecting various modern-day diseases including obesity, diabetes, inflammatory bowel diseases, and colorectal cancer.

SCFAs contribute up to 5%–10% of the total energy in a healthy body. Acetate in the blood plays a key metabolic role for peripheral tissues being a substrate for energy production, lipogenesis, and cholesterol synthesis. Propionate reduces energy intake by attenuating reward-based eating behavior in the striatum, the subcortical part of the forebrain that is a critical component of the reward system. Butyrate acts as an important energy and carbon source for colonocytes, is protective against colon cancer acts as an antiinflammatory, antioxidative agent, and plays a role in intestinal barrier functions. Butyrate has multiple effects on brain function and behavior. In the light of new findings regarding mechanisms of action of SCFAs in general and butyrate in particular, a hypothesis was put forth: a high fiber diet may improve brain health.

Bacteria found in the human gut synthesize and excrete vitamins in excess of their own needs, which we can absorb and use as nutrients. For example, gut bacteria secrete Vitamin K, Vitamin B12, and other B-vitamins. Moreover, germ-free animals and the human newborn are deficient in Vitamin K to the extent that it is necessary to supplement their diets or administer an injection at birth.

Microalgae are found in many marine invertebrates, including corals, mollusks, and sponges. These photosynthetic symbionts provide their hosts with nutrients. In the case of corals, the microalgae, commonly called zooxanthellae, provide more than 50% of the carbon and energy requirements of their hosts by producing sugars via photosynthesis and then transferring the sugar to the coral. Thus, the symbiosis of corals and zooxanthellae makes possible densely populated coral reefs in oligotrophic (nutrient-poor) marine environments. Symbiotic animals containing microalgae challenge the common perception that only plants can capture the sun's rays and converting them into biological energy through photosynthesis.

An unexpected symbiosis, referred to as "chemosynthetic symbiosis," between bacteria and marine invertebrates was discovered 35 years ago, at hydrothermal vents on the Galapagos Rift. Remarkably, it took the discovery of this symbiosis in the deep sea for scientists to realize that chemosynthetic symbioses occur worldwide in a wide range of habitats, including shallow-water coastal sediments, sea-grass beds, and wherever large organic food falls to the deep sea—such as whale carcasses and wood logs. In chemosynthetic symbioses, bacteria obtain energy by oxidizing inorganic material, such as hydrogen sulfide (H_2S) and use the energy to synthesize organic matter from carbon dioxide (CO_2). The organic matter they synthesize is the primary source of nutrition for their animal host. In turn, the animal host provides its symbionts a habitat in which they have access to the compounds they need, H_2S, O_2, and CO_2. Together, these partners create organisms with novel metabolic capabilities. The evolutionary success of these symbioses is evident from the wide range of animal groups that have established associations with chemosynthetic bacteria; at least seven animal phyla are known to host these symbionts. The fact that complex life, comprised of symbioses between chemosynthetic bacteria and animals, exists in the absence of light has widened the possibilities of extraterrestrial life.

Insects are the most diverse animal group on earth, embracing several million species. Most insects live on an unbalanced diet that requires supplementation from their microbiota. For example, aphids and other sap-feeding insects feed on a diet rich in sugar but poor in essential amino acids and vitamins. Other insects, such as the tsetse fly, feed solely on blood. In several cases, neither the host insect nor the bacterial symbiont can survive without the other (absolute mutualism). For example, in the symbiosis between the aphid and a bacterium called *Buchnera*, the bacterium is harbored in the abdominal body cavity of the aphid and provides essential amino acids that are lacking in the phloem sap diet of the insect. The aphid partner depends on these essential amino acids that are synthesized and furnished by the symbionts.

Termites are among a few animal species that can utilize wood as an energy source due to a hindgut containing highly specialized microbes. The hind gut of wood-feeding termites is a tiny but astonishingly efficient bioreactor, in which diverse microbes catalyze the conversion of plant cell walls (mainly cellulose) to glucose and fermentation products, mainly acetic acid that feed their host. DNA analyses have revealed the presence of hundreds of microbial species in this tiny environment.

The bovine rumen symbiosis has been studied extensively for many years, largely because of its obvious commercial importance. The rumen (first and largest division of the stomach in ruminant animals) acts as a temperature-controlled anaerobic fermenter that mainly processes ground cellulose-containing material together with saliva from the cow's mouth. In the rumen, bacterial enzymes convert the cellulose into its glucose subunits, which are then fermented by different groups of bacteria to produce short-chain fatty acids. These fatty acids are absorbed through the wall of the rumen into the bloodstream and then circulated via the blood to the various tissues of the body where they serve as nutrients. The microbial population of the rumen grows rapidly, and a mass of the microbial cells passes periodically out of the rumen with undigested plant material into the lower stomachs. There the protein-rich microbial mass is digested by secreted enzymes of the host, providing nitrogenous compounds and vitamins that are absorbed into the blood and used by the animal. The cow benefits from the cooperation by being able to grow and reproduce on a simple and abundant diet of cellulose, water, and inorganic salts, in spite of the fact that it lacks the ability to synthesize cellulases and some vitamins and essential amino acids. The concept that symbiotic microorganisms benefit their hosts by allowing them to derive energy from complex compounds and by providing essential nutrients is a general phenomenon in animals, including humans.

Also, plants obtain essential nutrients from their microbial symbionts. Most plant species enter into a mutualistic root symbiosis with fungi, in which plant sugar, primarily glucose, is traded for fungal minerals, mainly phosphorus and water, thereby protecting the plant against environmental stress, such as drought. This is an ancient symbiosis, which has been detected in fossils of early land plants. In addition to fungi, many bacterial species interact with plant roots and contribute to carbon transfer to soil, nitrogen-fixation, mineralization of organic materials, solubilization of minerals in rocks by excretion of organic acids, maintenance of soil structure, and water cycling, all of which promote plant growth directly or indirectly. Beneficial plant bacteria are sometimes referred to as plant-growth promoting bacteria.

Several specialized kinds of bacteria, including the most studied, *Rhizobium*, engage in symbiotic relationships with peas, soybeans, and other legumes to convert nitrogen gas (N_2) into ammonia (NH_3) and further into organic nitrogen-containing compounds. This fixed nitrogen is subsequently assimilated by the host plant, resulting in improved growth and productivity, even under N-limiting environmental conditions. Farmers use soybeans and other legumes in rotations with grass crops, such as corn or wheat. Grass crops are unable to take their own nitrogen from the air so they either need the nitrogen in the soil that the legumes provide for them in a crop rotation or they need a chemical fertilizer containing nitrogen. In a 2-year rotation, a farmer will alternate a year of a legume, such as soybeans and a year of a grass crop, such as corn. Symbiotically fixed N accounts for 90 million metric tons per year, or nearly half the annual quantity of nitrogen entering the soil.

Microbiotas influence human, animal, and plant development. Developmental biology is the branch of biology dealing with the processes of growth and change that transform an organism from a fertilized egg or asexual reproductive unit to an adult. In recent years, a number of exciting experiments have demonstrated that microbial symbionts contribute to development programming of a variety of tissues, functions, and organs.

The symbiosis between the Hawaiian bobtail squid *Euprymna scolopes* and the luminous (light-producing) bacterium *Vibrio fischeri* is one of best studied systems that demonstrate how a bacterial symbiont can play a role in the development of an animal organ. The bacteria do not produce light when they are freely roaming in the ocean, but when housed in the squid's light organ (located in its underbelly) it will interact with the animal to emit light according to how much moonlight and sunlight is visible above. In doing so, the squid will glow a light blue to mimic the light from above, eliminating its shadow on the seabed and rendering it invisible to predators potentially lurking below. The two live a happy coexistence: the bacteria getting sustenance from the squid, the squid getting camouflage from the bacteria. Development of the light organ is controlled by the bacterium. Following fertilization of the eggs within the female, the embryos develop an immature light organ that is free of bacteria. The female host lays clutches of hundreds of fertilized eggs, which hatch almost synchronously at dusk. Within hours after hatching, the juvenile squid becomes colonized by the *Vibrio*, which triggers changes in the light organ. The presence of as little as five *Vibrio* cells in the immature light organ for 12 h is sufficient to induce development of the mature light organ. In the absence of the specific symbiont, the light organ does not develop.

Development of an efficient immune system is crucial for the health and survival of all organisms. The immune system can be subdivided into two categories: innate and adaptive. Innate immunity (meaning "present from birth") refers to nonspecific defense mechanisms, such as the skin, the flushing action of tears and saliva, skin oils, antibacterial biochemicals in the blood and tissues, and immune cells that nonspecifically attack foreign cells and materials in the body. The normal microbiota of all animals and plants should be considered part of the innate immune system because, as discussed earlier, they are a first line of defense against pathogens.

The microbiota also primes development of the adaptive or antigen-specific immune system. The adaptive immune system is exactly that—it's adaptive, meaning it can adapt to a specific threat, or antigen. An antigen (short for antibody generator) is a foreign substance that induces an immune response in the body, especially the production of antibodies. The adaptive immune response is more complex than the innate. The antigen first must be processed and recognized. Once an antigen has been recognized, the adaptive immune system creates an army of immune cells specifically designed to attack that antigen. Adaptive immunity also includes a "memory" that makes future responses against a specific antigen more efficient. This means that whenever they encounter that specific antigen or threat, again, they will know exactly what to do with it and will attack it immediately. This is the basis of immunization.

Throughout life, gut bacteria shape the tissues, cells, and molecular profile of the mammalian immune system. This partnership is based on a molecular exchange involving bacterial signals that are recognized by host receptors to mediate beneficial outcomes for both microbes and hosts. Evidence shows that the resident microbiota "programs" many aspects of immune cell differentiation, thus augmenting the developmental instructions of the host genome to engender the full function of the adaptive immune system.

Germ-free mice exhibit significant differences in gut development and function as compared with mice possessing normal gut microbiota. The germ-free mice demonstrate enlarged caeca (pouch forming the beginning of the large intestine), a slow digested food transit time, and altered rate of cell turn-over in the small intestine. The immune system distinguishes between self and foreign antigens and mounts an appropriate response to clear invading pathogens. It is likely that the immune system first evolved to preserve and protect its own microbiota and subsequently was used to kill dangerous foreign microorganisms.

Indole acetic acid (IAA) is the most common as well as the most studied plant hormone, or auxin. IAA affects plant cell division, extension, and differentiation, stimulates seed and tuber germination, increases the rate of xylem and root development, controls processes of vegetative growth, and initiates lateral, and adventitious root formation. Most plant-stimulating bacteria produce IAA, including those that form specific symbiotic relationships with plants. IAA synthesized by bacteria may be involved at different levels in plant-bacterial interactions. In particular, IAA produced by bacteria in the rhizosphere play a major role in root development.

Microbiotas influence human and animal behavior. Mice experiments have demonstrated that gut microbiota affects the brain and behavior. Germ-free mice are more active and spend more time scurrying around their enclosures than conventional mice. They are also less anxious and more likely to take risks, such as spending long periods of time in bright light or open spaces, compared to the normal mice. Inoculating gut microbiota from healthy mice into germ-free baby mice caused them to behave in the "normal" cautious way. However, if germ-free adult mice are inoculated with gut bacteria, their behavior does not change, suggesting that the microbiota affects the early development of the brain that subsequently influences adult behavior. There appears to be a critical window during development when microbiota influences the central nervous system wiring related to stress-related behaviors.

There is a large difference in how genes are expressed (turned on and off) in the brain between germ-free and conventional mice. Some of these genes are involved in providing cells with energy, others in chemical communications across the brain, and yet others in strengthening the connections between nerve cells. The data suggest that during evolution, the colonization of gut microbiota has become integrated into the programming of brain development, affecting motor control and anxiety-like behavior.

How do gut bacteria affect the brain? To begin with, the long cranial vagus nerve passes through the neck and thorax to the abdomen and transmits information about what happens in the gut to the brain and vice versa. But the bacteria also signal the brain via changing levels of dietary components and hormones. Hormones, by definition, can affect parts of the body over long distances. For example, plasma levels of the neurotransmitter serotonin are 2.8-fold higher in conventional mice than germ-free animals. Since bacteria do not synthesize serotonin, it is likely that the increased level of plasma serotonin results from an as yet undefined host microbe interaction.

With regard to physical and psychological stress, the interaction of gut bacteria with the brain is bidirectional. Stress has been shown to affect the composition of intestinal microbiota in rodents and primates, and as was discussed earlier commensal microbes affect the neural network responsible for controlling stress responsiveness. There is also growing evidence that gut microbiota affects autism. Autism, a developmental disorder that causes impediments to social interactions and behavior, is usually linked by scientists to abnormalities in brain structure and function, caused by a mixture of genetic and environmental factors. However, recent evidence suggests a link between autism and gut microbiota. Researchers from the Baylor College of Medicine, Texas, found that the presence of a single species of gut bacteria isolated from the human gut and placed in mice could reverse many behavioral characteristics related to autism. Interestingly, autism and accompanying gastrointestinal (GI) symptoms are characterized by distinct and less diverse gut microbial compositions. One case report describes the benefits of antimicrobial therapy on behavior in a 14-year old boy with autism. Over the course of treatment, the boy exhibited a reduction in aberrant behaviors, increased gastrointestinal function, and improved quality of life. The fact that antibiotics relieved the symptoms further suggests a connection between bacteria and autism. The data suggest that probiotic intervention should be explored for the treatment of neurodevelopmental disorders in humans.

It has been argued that access to mutualistic symbiotic microbes is an under-appreciated benefit of group living and was probably a strong selective force in the evolution of social behavior. Kissing, hugging, and touching ensure that offspring acquire the beneficial microbiota of the group. This concept can be expanded by suggesting that gut microbiota plays a role in kin recognition, the cognitive process by which animals distinguish kin and nonkin. This is an important biological property because the ability to recognize one's relatives provides a mechanism allowing for the emergence of sociality. Not only kin, but nest mates or group members, who share microbiota, can be recognized by their common microbial-determined odors.

Obesity. During the last few years there have been numerous reports in the scientific literature and popular press claiming that bacteria play an important role in obesity. The initial hint came from the observation that obese and lean animals (mice and humans) have vastly different microbiotas in their guts and that obese individuals had fewer types of bacteria. People whose guts contain fewer types of bacteria and bacterial genes were found to contain higher levels of body fat and inflammation than those with more different types of gut bacteria.

An elegant experiment performed in the laboratory of Jeffrey Gordon at the Washington University School of Medicine in St. Louis provided strong support for the bacterial hypothesis of obesity. One group of germ-free mice was inoculated with bacteria from obese humans and the second group with bacteria from lean humans. In spite of the fact that both groups of mice were given precisely the same amount of food, the mice that received "obese bacteria" gained much more weight than the mice that received "lean bacteria." The team also showed that a lean microbial community could infiltrate and displace an obese one, preventing mice from gaining weight so long as they were on a healthy diet. These data indicate a relationship between obesity and the gut microbiota and open the exciting future possibility of treating obesity by gut microbiota manipulation.

Although obesity is now defined by the World Health Organization as a disease, caution should be applied in considering "obese bacteria" as pathogens and trying to eliminate them for the following reason: during the third trimester of a normal pregnancy, women acquire a gut microbiota which is similar to that found in obesity. These same "obese bacteria" may be highly beneficial during pregnancy, as they promote energy storage in fat tissue and provide for the growth and development of the fetus, which is central to the reproduction of mammalian species.

Bacteria play a role in mating preference. Microbes are largely responsible for the odor of animals, and odor plays an important role in mating preference in many animals including humans. It has been established that symbiotic bacteria play an essential role in determining the unique odors of several animals, including fish, rats, deer, bats, and humans. Microorganisms contribute to the odor of animals by producing volatile short-chain fatty acids, alcohols, and ketones which are prominent and active components of mammalian scent. Also, microbes convert the odorless organic products of sweat glands to a variety of compounds, such as steroids, sulfanyl alkanols, and branched fatty acids that yield the characteristic odors of specific individuals. The process of odor production is complex, and requires multiple bacterial species.

In 1989, Diane Dodd, a graduate student at Yale University at the time, published a remarkable observation that could not be explained. She divided a population of fruit flies into two parts: one part was fed a starch diet, the other part a simple sugar diet. After about a year (18 generations), she mixed the two populations. What she observed was that the "starch flies" tended to mate with "starch flies" and "sugar flies" tended to mate with "sugar flies." The result was surprising since according to existing theory the time was too short to allow the fly genomes to change by mutation. Although unexplained, the data were considered important because mating selection is an early stage in the emergence of new species, and the origin of species is a hallmark of Darwinian evolution.

FIGURE 13.2 Symbiotic bacteria are largely responsible for the odor of animals and mating preferences.

Dodd's results only became comprehensible in 2010 when Gil Sharon, a graduate student in my laboratory, showed that changing the diet of fruit flies causes a rapid multiplication of a particular bacterial symbiont, *Lactobacillus plantarum*, and that this bacterium is responsible for the mating preference. The symbiotic *L. plantarum* influences mating preference by changing the levels of sex pheromones produced by the fly. Thus, microbiotas can play a role in mating preference. In nature, a combination of partial geographic separation and bacterial-induced mating preference would reduce interbreeding of populations. Slower changes in the host genome would further enhance the mating preference. The stronger the mating preference, the greater the chance that two populations will become sexually isolated, and evolutionary biologists have argued that the emergence of sexual isolation is the central event in the origin of species (Fig. 13.2).

In conclusion, it is now clear that the microbiome with its enormous DNA information plays an important part in the health of all animals and plants. Protection against pathogens is one of the most general and important contributions of resident microbiota to the health of organisms. Provision of essential nutrients is another general benefaction of microbiota to their hosts. In humans, the gut microbiota is a complex ecosystem that plays an essential role in the breakdown of dietary fibers, production of vitamins and amino acids, detoxification of harmful chemicals, regulation of angiogenesis and blood pressure, and development of the innate and antigen-specific immune systems. Although the mechanisms are not clear, gut bacteria in humans and mice influence obesity and affect the brain.

In the next two chapters, I will present current hypotheses on the origin of the first cells and how both host and microbial DNA information have contributed to the evolution of biological complexity.

NOTES AND REFERENCES

116 Pioneers in research on the human microbiome include Jeffrey Gordon at the Washington School of Medicine in St. Louis, Martin Blaser at the New York University School of Medicine and Forest Rohwer of San Diego State University.

116 Gillespie, R.G., Roderick, G.K., 2002. Arthropods on islands: colonization, speciation, and conservation. Annu. Rev. Entomol. 47, 595–632.

116 Shanmugam, M., et al., 2005. Bacterial-induced inflammation in germ-free rabbit appendix. Inflamm. Bowel Dis. 11, 992–996.

116 Silva, A.M., et al., 2004. Effect of *Bifidobacterium longum* ingestion on experimental salmonellosis in mice. J. Appl. Microbiol. 97, 29–37.

116 Huppert, M., et al., 1955. Pathogenesis of *Candida albicans* infection following antibiotic therapy. J. Bacteriol. 70, 4356–4439.

116 Khoruts, A., Sadowsky, M.J., 2011. Therapeutic transplantation of the distal gut microbiota. Mucosal Immunol. 4, 4–7.

116 Gough, E., et al., 2011. Systematic review of intestinal microbiota transplantation (fecal bacteriotherapy) for recurrent *Clostridium difficile* infection. Clin. Infect. Dis. 53, 994–1002.

119 Stilling, R.M., van de Wouw, M., Clarke, G., Stanton, C., Dinan, T.G., Cryan, J.F., 2016. The neuropharmacology of butyrate: the bread and butter of the microbiota-gut-brain axis? Neurochem. Int. 99, 110–132.

119 Rumpho, M.E., et al., 2011. The making of a photosynthetic animal. J. Exp. Biol. 214, 303–311.

120 Ponsard, J., et al., 2013. Inorganic carbon fixation by chemosynthetic ectosymbionts and nutritional transfers to the vent host-shrimp *Rimicaris exoculata*. ISME J. 7, 96–109.

120 Wilson, A.C.C., et al., 2010. Genomic insight into the amino acid relations of the pea aphid *Acyrthosiphon pisum* with its symbiotic bacterium *Buchnera aphidicola*. Insect Mol. Biol. 19, 249–258.

120 Brune, A., 2011. Microbial symbioses in the digestive tract of lower termites. In: Rosenberg, E., Gophna, U. (Eds.), Beneficial Microorganisms in Multicellular Life Forms. Springer, Heidelberg.

120 Mizrahi, I., 2011. Role of the rumen microbiota in determining the feed efficiency of dairy cows. In: Rosenberg, E., Gophna, U. (Eds.), Beneficial Microorganisms in Multicellular Life Forms. Springer, Heidelberg, Germany.

121 Sekirov, I., et al., 2010. Gut microbiota in health and disease. Physiol. Rev. 90, 859–904.

121 Pluznicka, J.L., et al., 2013. Olfactory receptor responding to gut microbiota derived signals plays a role in renin secretion and blood pressure regulation. Proc. Natl. Acad. Sci. USA 110, 4410–4415.

121 Bloemberg, G.V., Lugtenberg, B.J., 2001. Molecular basis of plant growth promotion and biocontrol by rhizobacteria. Curr. Opin. Plant Biol. 4, 343–350.

122 Rawis, J.F., et al., 2004. Gnotobiotic zebrafish reveal evolutionary conserved responses to the gut microbiota. Proc. Natl. Acad. Sci. USA 101, 4596–4601.

122 Nyholm, S.V., McFall-Ngai, M., 2004. The winnowing: establishing the squid vibrio symbiosis. Nat. Rev. Microbiol. 2, 632–642.

122 Lee, Y.K., Mazmanian, S.K., 2010. Has the microbiota played a critical role in the evolution of the adaptive immune system? Science 330, 1768–1773.

122 Ridaura, V.K., et al., 2013. Gut microbiota from twins discordant for obesity modulate metabolism in mice. Science 341, 1241214.
122 Walker, A.W., Parkhill, J., 2013. Fighting obesity with bacteria. Science 6 (341), 1069–1070.
122 Koren, O., et al., 2012. Host remodeling of the gut microbiome and metabolic changes during pregnancy. Cell 150, 470–480.
123 Heijtz, R.D., et al., 2011. Normal gut microbiota modulates brain development and behaviour. Proc. Natl. Acad. Sci. USA 108, 3047–3052.
124 Gonzalez, A., et al., 2011. The mind-body-microbial continuum. Dialog. Clin. Neurosci. 13, 55–62.
124 Ramirez, P.L., et al., 2013. Improvements in behavioral symptoms following antibiotic therapy in a 14-year-old male with autism. Case Rep. Psychiatry, 239034.
124 Lombardo, M., 2008. Access to mutualistic endosymbiotic microbes: an underappreciated benefit of group living. Behav. Ecol. Sociobiol. 62, 479–497.
124 Archie, E.A., Theis, K.R., 2011. Animal behaviour meets microbial ecology. Anim. Behav. 82, 425–436.
124 Shawkey, M.D., et al., 2007. Bacteria as an agent for change in structural plumage color: correlational and experimental evidence. Am. Nat. 169, 112–121.
125 Dodd, D.M.B., 1989. Reproductive isolation as a consequence of adaptive divergence of *Drosophila pseudoobscura*. Evolution 43, 1308–1311.
125 Sharon, G., et al., 2010. Commensal bacteria play a role in mating preference of *Drosophila melanogaster*. Proc. Natl. Acad. Sci. USA 107, 20051–20056.
125 Swann, J., et al., 2009. Gut microbiome modulates the toxicity of hydrazine: a metabonomic study. Mol. Biosyst. 5, 351–355.
125 Monachesea, M., et al., 2012. Bioremediation and tolerance of humans to heavy metals through microbial processes: a potential role for probiotics? Appl. Environ. Microbiol. 78, 6397–6404.
125 Kuz'mina, V.V., Pervushina, K.A., 2003. The role of proteinases of the enteral microbiota in temperature adaptation of fish and helminthes. Doklady Biol. Sci. 391, 2326–2328.
125 Rodriguez, R., Redman, R., 2008. More than 400 million years of evolution and some plants still can't make it on their own: plant stress tolerance via fungal symbiosis. J. Exp. Biol. 59, 1109–1114.

Chapter 14

Origin of Nucleic Acids and the First Cells

An honest man, armed with all the knowledge available to us now, could only state that in some sense, the origin of life appears at the moment to be almost a miracle, so many are the conditions which would have had to have been satisfied to get it going.

—Francis Crick, *Life Itself: Its Origin and Nature*

The uniformity of the earth's life, more astonishing than its diversity, is accountable by the high probability that we derived, originally, from some single cell, fertilized in a bolt of lightning as the earth cooled. It is from the progeny of this parent cell that we take our looks; we still share genes around, and the resemblance of the enzymes of grasses to those of whales is a family resemblance.

—Lewis Thomas, *The Lives of a Cell: Notes of a Biology Watcher*

The fundamental desire to understand the origin of life is demonstrated by the fact that essentially all human cultures contain a story of how life began. From the tribes of ancient times to the mythologies of more modern cultures, there are countless stories of the origin of life. Some are based in pagan beliefs, while others are based on creation resulting from a holy deity. This collection of nativities, myths, legends, and tribal knowledge handed down over generations is the collective expression of how man has attempted to explain his world and his place in it. What is common to all these cultures is that their specific story of how life began is accepted without question. The role of the adults is simply to teach the story to the children.

The distinguishing feature of the scientific approach to understanding the origin of life is that it begins by acknowledging the fact that we do not know how life began. Only by admitting that we do not have the answer, is it possible to put forth and test multiple hypotheses. The origin of life and nucleic acids remains one of the most challenging problems in biology.

In discussing the origin of life, the first experimental finding that has to be considered is that prokaryotic life (microorganisms lacking a membrane bound nucleus, such as bacteria and cyanobacteria) first appeared on Earth about 3.7 billion years ago. The evidence for this is fossil remains of cyanobacteria found in ancient rock formations (stromatolites) in Western Australia and

It's in Your DNA. http://dx.doi.org/10.1016/B978-0-12-812502-1.00014-7

Greenland. Since the Earth's surface cooled sufficiently to allow for life only about 4.1 billion years ago, it is reasonable to assume that prokaryotic life originated sometime between 3.7 and 4.1 billion years ago. Stromatolites would have to be the descendants of the earlier life-forms. Cyanobacteria capable of performing photosynthesis and forming communities are relatively sophisticated organisms. They presumably had less-sophisticated ancestors that lived more than 3.7 billion years ago. During this time, the Earth's atmosphere consisted of methane (CH_4), hydrogen (H_2), water (H_2O), nitrogen (N_2), and ammonia (NH_3). Note the absence of oxygen (O_2). Recent evidence suggests that carbon dioxide (CO_2) and sulfur dioxide (SO_2) were also present in low amounts in the primitive atmosphere.

In 1936, a Russian biochemist, Alexander Oparin, published a book, *The Origin of Life on Earth*, which revolutionized thinking on the subject and provided a new experimental approach to the problem. Oparin's theory is fundamentally different from previous theories because it requires no special laws to explain the origin of life. The phrase "chemical evolution" is used to emphasize the gradual appearance of living things. Much like Darwin's story of the *Origin of Species*, Oparin postulates a long series of chemical changes as a prerequisite to the formation of life. Both Darwin and Oparin suggest a gradual transition from simple to more complex structures. Oparin's theory of chemical evolution leading to the origination of life can be considered in four phases:

$$\text{Inorganic gases} \rightarrow \text{Organic molecules} \rightarrow \text{Polymers} \rightarrow \text{Aggregates} \rightarrow \text{Protocells}$$

Oparin's abiogenesis theory of the origin of life begins with the concept that simple organic compounds, such as amino acids, organic acids, purines, pyrimidines, and sugars, were produced spontaneously in the earth's primitive oxygen-free atmosphere. Experimental support for this concept was published in 1953, when a graduate student, Stanley Miller and his mentor the Nobel Prize chemist Harold Urey at the University of Chicago, showed that a variety of small organic compounds were, in fact, produced when an electric spark, mimicking an electric storm, was passed through a mixture of the primitive atmospheric gases. Since that time many researchers have obtained similar results using different energy sources, such as heat, ultraviolet light, and ionizing radiation. Thus, the first phase of Oparin's theory has a solid experimental basis. Small molecules, such as amino acids, sugars, purines, and pyrimidines, could and should have been present in the primitive oceans (Fig. 14.1).

Whatever the earliest events on the road to the first living cell were, it is clear that at some point the large biological molecules found in modern cells must have emerged. Considerable debate in origin-of-life studies has revolved around the question: Which of the fundamental macromolecules appeared first, proteins, DNA, or RNA? To gain insight into this question, it is useful to consider the functions performed by each of these large molecules in existing organisms.

FIGURE 14.1 **The Miller–Urey experiment—better than expected.**

The proteins make up about 50% of the mass of most cells and are the main structural and functional agents in the cell. Catalytic proteins, or enzymes, carry out thousands of chemical reactions that take place in any given cell, among them the synthesis of all other biological constituents, including DNA and RNA. However, proteins cannot replicate themselves. They require the information contained within the nucleic acids, DNA, and RNA. In all modern cellular organisms, DNA serves as the storage site of genetic information. In viruses, either DNA or RNA is used to store genetic information. The nucleic acids contain the instructions for the manufacture of proteins. In the modern cell, *protein, DNA, and RNA are each dependent on the others for their manufacture and function.* DNA, for example, is merely a blueprint, and cannot perform a single catalytic function, nor can it replicate on its own. Proteins, on the other hand, perform most of the catalytic functions, but cannot be manufactured without the specifications encoded in DNA and transcribed to RNA. This classic "chicken-and-egg" problem makes it immensely difficult to conceive of any plausible prebiotic chemical pathway to the current biological system.

One possible scenario for life's origins is that two kinds of macromolecules evolved together, one informational and one catalytic. But scientists studying the origin of life consider this scenario extremely complicated and highly unlikely. Another possibility, currently favored by many theorists, is that one of these molecules, namely RNA, could itself perform the multiple functions of self-replication and catalysis. Catalytic RNAs, also referred to as ribozymes (*ribo*nucleic acid en*zymes*), were discovered in the 1980s. Today, RNA molecules are the only molecules known both to store genetic information (as in the RNA viruses and in the form of messenger RNA) and to exert biological catalysis, like a protein enzyme. RNA may therefore have supported precellular life and been a major step in the evolution of cellular life. This hypothesis, referred to as the "RNA world" by Walter Gilbert of Harvard University, has gained support in recent years from experiments demonstrating that ribozymes can catalyze the cleavage of messenger RNA molecules and can also catalyze their own synthesis under very specific conditions.

The RNA world hypothesis assumes a phase of life, whereby catalytic biopolymers consisted exclusively of ribozymes, hence the name "RNA world." The hypothesis assumes that translation of RNA sequences into protein sequences began with the appearance of a primitive form of a ribosome. This gave rise to an "RNA-protein world" with a step-by-step replacement of ribozymes by enzymes (enzymatic take-over). Much later, according to this theory, RNA as the replicating informational molecule was replaced by the chemically more stable double-stranded DNA. Although still highly speculative and criticized by several scientists, the RNA world hypothesis is considered the most promising concept we currently have to help understand the backstory to contemporary biology.

Another important issue to consider in the origin of life is boundaries, how molecules might have assembled into the first cell-like structures, or "protocells." The origin of cell membranes is a major unresolved issue in the Oparin theory of the origin of life. Contemporary cells are enclosed by membranes that are made from lipids into which different proteins are embedded. Membranes keep the cell components physically together and form a barrier to the uncontrolled passage of large molecules. Specialized proteins embedded in the membrane act as gatekeepers and pump molecules in and out of the cell. We do not know how a rudimentary protocell could carry out this task?

I would like to emphasize that attempts to explain the origin of life are rich in speculation and poor on evidence. When a particular concept is tested in a laboratory and yields a result consistent with the hypothesis, then the hypothesis moves from pure speculation to a possible hypothesis. For example, the formation of amino acids from gases in the Earth's prebiotic atmosphere is a possible step in the origin of life because it has been experimentally demonstrated. The RNA world hypothesis is still primarily speculation.

An alternative to the concept that life originated on Earth by a series of prebiotic chemical reactions is the Panspermia Theory. The word "Panspermia" comes from the Greek language and means "seeds everywhere." The seeds in this case would not only be the building blocks of life, such as amino acids and sugars, but also small heat-resistant microorganisms. The theory states that these "seeds" were dispersed "everywhere" from outer space and most likely came to Earth from meteorite impacts. Once the earth had cooled sufficiently, these invading microbes found the conditions favorable for growth and gave rise to life on this planet.

Recent research has provided some support for the Panspermia Theory. First, we now know that some microbes are resistant to extreme conditions and may be able to survive for very long periods of time, probably even in deep space and, hypothetically, could travel in a dormant state between hospitable environments. Second, researchers have reported the presence of complex organic molecules and possibly extraterrestrial bacteria inside meteorites. Third, water-rich meteoroids, colliding with the Earth, most likely brought water to the world's oceans. If so, it certainly could have brought microbes.

The Panspermia Theory has been criticized because it dodges the question. If the Earth was infected by microbes, we still have to explain how these microbes came into existence on their native planet. Nevertheless, the Panspermia Theory has been supported by several leading scientists, including the theoretical physicist and cosmologist Stephen Hawking, Nobel Prize winner Francis Crick, and the chemist and evolutionary biologist Leslie Orgel. The panspermiatic premise of a meteor bringing alien life to Earth is the basis of much science fiction, including Jack Finney's novel *The Body Snatchers*, Ursula K. Le Guin's series the *Hainish Cycle*, Michael Crichton's novel *The Andromeda Strain*, Stephen King's short story *Weeds*, and in *Star Trek: The Next Generation* episode, *"The Chase."*

Prokaryotes (Greek for "before *karyon*" or "before nucleus") are simple, single-cell organisms that lack a membrane-bound nucleus. Eukaryotes (Greek for "true nucleus") possess a membrane-bound nucleus, an intricate cytoskeleton, and membrane-bound structures in the cytoplasm, such as mitochondria, where most of the cell's energy is produced, and in the case of algae and plants cells, also chloroplasts, which are the sites of photosynthesis. Based on the fossil record, eukaryotes (filamentous algae) first appeared on Earth about 1.5 billion years ago. Since prokaryotes were present 3.7 billion years ago, it follows that prokaryotes were the only cellular form of life on Earth for more than 2 billion years. During this time, prokaryotes evolved most of the biochemical reactions present in all forms of life, including DNA replication, the genetic code, protein synthesis via transcription and translation, photosynthesis, and the synthesis and breakdown of most organic compounds under anaerobic and aerobic conditions. During this time, prokaryotes split into two of the three groups, or domains of life, called Bacteria (or eubacteria) and Archaea (or archaebacteria). The third domain of life, Eukarya, contains all the eukaryotic organisms: animals, plants, and single-celled organisms that contain a true nucleus, such as algae, yeasts, and other fungi.

Since eukaryotic organisms, including humans, perform the same basic biochemical reactions as prokaryotes, it is highly likely that eukaryotes descended from prokaryotes. In support of this hypothesis is the fact that 60% of the approximately 22,000 human genes that code for proteins are similar to prokaryotic genes. But how did the eukaryote cell evolve from the prokaryote cell? For many years scientists assumed that the transition from prokaryote to eukaryote occurred through the familiar process of mutation and natural selection. However, in 1971, Lynn Margulis at the University of Massachusetts at Amherst, published her book *The Origin of Eukaryotic Cells*, which posited that a number of parts of the eukaryote cell were acquired in a radically different way, namely by the fusion of separate bacterial species. This revolutionary concept is referred to as the "endosymbiont hypothesis." As Margulis wrote:

> *How did the eukaryotic cell appear? Probably it was an invasion of predators, at the outset. It may have started when one sort of squirming bacterium invaded another—seeking food, of course. But certain invasions evolved into truces; associations once ferocious became benign. When swimming bacterial would-be invaders took up residence inside their sluggish hosts, this joining of forces created a new whole that was, in effect, far greater than the sum of its parts: faster swimmers capable of moving huge quantities of genes evolved. Some of these newcomers were uniquely competent in the evolutionary struggle. Further bacterial associations were added on, as the modern cell evolved.*

Lynn Margulis (born Lynn Petra Alexander) was an exceptionally gifted woman. She attended the University of Chicago at age 14 having entered "because I wanted to go and they let me in." At 19, she married the astronomer Carl Sagan. Their marriage lasted 8 years. Later, she married Dr. Thomas Margulis, a crystallographer. Lynn Margulis was a strong critic of Charles Darwin's gradual selection theory and modern evolutionary theory. She stated: "Although I greatly admire Darwin's contributions and agree with most of his theoretical analysis and I am a Darwinist, I am not a Neo-Darwinist." Her position sparked a lifelong debate with leading Darwinian biologists, including Richard Dawkins, George C. Williams, and John Maynard Smith. Basically, Margulis argued that random mutation was too slow to account for evolution. She died in 2011 at the age of 73 of a hemorrhagic stroke (Fig. 14.2). In 1995, the English evolutionary biologist Richard Dawkins had this to say about Lynn Margulis and her work:

> *I greatly admire Lynn Margulis's sheer courage and stamina in sticking by the endosymbiosis theory, and carrying it through from being unorthodoxy to orthodoxy. I'm referring to the theory that the eukaryotic cell is a symbiotic union of primitive prokaryotic cells. This is one of the great achievements of twentieth-century evolutionary biology, and I greatly admire her for it.*

Many studies have bolstered this once-controversial endosymbiont hypothesis. Let us first consider the mitochondrion, a fundamental organelle of all eukaryotic cells. Mitochondria resemble bacteria in many ways. Both are of similar size and are surrounded by a membrane. Mitochondria and bacteria can use oxygen to generate chemical energy, in the form of adenosine triphosphate (ATP) molecules. Mitochondria have their own bacterial-like DNA, which they duplicate when they divide, similarly to bacteria, into new mitochondria. Based on DNA analyses, mitochondria are closely related to the intracellular parasite

FIGURE 14.2 **Lynn Margulis (1938–2011).**

Rickettsia prowazekii, the causative agent of epidemic typhus. The ribosome coded for by mitochondrial DNA is similar to those from bacteria in size and structure.

Using similar arguments, it is now generally accepted that chloroplasts of photosynthetic algae and plants arose through incorporation into cells of symbiotic cyanobacteria. It is thought that photosynthesis first evolved in a group of bacteria known as cyanobacteria about 2.5 billion years ago. Then, roughly 1 billion years ago, a single-celled ancestor of plants and algae engulfed a cyanobacterium. This led a symbiotic relationship, with the smaller cyanobacterial cell providing sugars by photosynthesis, and the larger cell providing other molecules that the cyanobacterium needed. Over time the two grew more and more attached to each other. They exchanged genetic information and became inseparable so that they essentially became a new type of photosynthetic organism. Over the course of the next billion years this organism evolved into the algae and plants we know today.

The fundamental difference between eukaryotes and prokaryotes is the presence of a membrane-bound nucleus in eukaryotes. Because the nucleus lacks an obvious homologue or precursor among prokaryotes, ideas about its evolutionary origin are diverse and highly speculative. One hypothesis is that a prokaryotic cell membrane formed an invagination that enclosed the DNA in a primitive prokaryotic cell and this membrane-bound DNA evolved into the nucleus. Another hypothesis is endosymbiosis of an archaebacterial cell within a eubacterial cell, with the archaebacterium becoming a nucleus. The reason for assuming that the archaebacterium became the nucleus is that the molecular machinery involved in information storage and retrieval in eukaryotes is more similar to archaebacterial counterparts than to eubacterial counterparts.

A third provocative option for the origin of the nucleus revolves around viruses. It has been proposed that a complex DNA virus became established in a prokaryote and evolved into the eukaryotic nucleus. The eukaryotic nucleus shares several properties with certain viruses, such as linear chromosomes and separation of transcription and translation. According to this hypothesis, a large virus would have taken control over a bacterial cell. Instead of replicating and destroying the host cell, it would enter into a symbiotic relationship with the cell. With the virus in control of the host cell's molecular machinery it would effectively become a "nucleus" of sorts. The hypothesis that a virus was the origin of the eukaryotic nucleus brings forth the question of the origin of viruses.

Since their discovery in the late 19th century, viruses have challenged our concept of what "living" means. Initially seen as toxic agents that cause disease, then as life-forms that multiply only in cells, and then as biological chemicals that could be crystallized, viruses are currently thought of as being in a gray area between living and nonliving. The Nobel laureate André Lwoff wrote, "Whether or not viruses should be regarded as organisms is a matter of taste. A virus is a virus." Three main hypotheses have been put forth to explain the origin of viruses:

1. Viruses arose from intracellular genetic elements that gained the ability to exit one cell and enter another.
2. Viruses are remnants of cellular organisms. Existing viruses may have evolved from more complex, possibly free-living organisms that lost genetic information over time, as they adapted to a parasitic approach to replication.
3. Viruses predate or coevolved with their current cellular hosts.

It should be pointed out that these hypotheses are not mutually exclusive. To distinguish between these hypotheses it would be useful to date the origin of viruses. Unfortunately, it has not yet been possible to detect ancient fossilized viruses. However, there is strong circumstantial evidence that viruses emerged very early in the evolution of life, before the separation of prokaryotes into two domains: Bacteria and Archaea. The reason for suggesting that the viruses arose before the separation of bacteria in two domains is that Bacteria and Archaea have the same types of viruses. It has been suggested that the first viruses were RNA viruses that originated during the RNA world and played a critical role in major evolutionary transitions, such as the invention of DNA.

Before discussing the evolution of animals and plants in the next chapter, I would like to summarize current hypotheses on the origin of the first cells. Life, in the form of prokaryotes, first appeared on earth about 3.7 billion years ago. Oparin's theory of the origin of life begins with the experimentally-supported concept that simple organic compounds would have been produced spontaneously in the earth's primitive reducing atmosphere. These molecules could have been concentrated into a "prebiotic organic soup," which by chemical evolution gave rise to larger molecules and eventually protocells. The Panspermia Theory considers that once the earth cooled it was infected with heat-resistant organisms from other celestial bodies.

The "RNA world" hypothesis assumes an early phase of life, wherein catalytic biopolymers consisted exclusively of ribozymes. Subsequently, a primitive form of a ribosome began the translation of RNA sequences into protein sequences. This gave rise to an "RNA-protein world." Much later, RNA as a replicating informational molecule was replaced by the chemically more stable DNA.

Single eukaryotic cells first appeared approximately 1.5 billion years ago, so that prokaryotes were the only cellular form of life for more than 2.2 billion years. During this time, prokaryotes evolved most of the biochemistry present in all forms of life and diversified into two groups, or domains, termed Bacteria (or eubacteria) and Archaea (or archaebacteria). The formation of eukaryotes involved the endosymbiosis of one prokaryote into another. Mitochondria were derived from a *Rickettsia*-like bacterium and chloroplasts from cyanobacteria. Concepts of the origin of the membrane-bound nucleus that characterizes eukaryotic cells are highly speculative. Hypotheses include an invagination of the prokaryote cell membrane to enclose the DNA, endosymbiosis of an archaebacterium within a eubacterium, with the archaebacterium becoming a nucleus, and a membrane-bound virus becoming established in a prokaryote and evolving into the eukaryotic nucleus by acquiring genes from the host. Although the origin of viruses remains an unsolved mystery, it is highly likely that virus-like entities appeared early in evolution.

The next chapter discusses the evolution of plants, animals, and humans.

NOTES AND REFERENCES

130 Fry, I., 2006. The origins of research into the origins of life. Endeavour 30, 24–28.
Zimmer, C., 2009. On the origin of eukaryotes. Science 325, 666–678.

130 The English chemist J.B.S. Haldane proposed a theory similar to Oparin's, which he arrived at independently, but a few years later.

131 Cech, T.R., et al., 1981. In vitro splicing of the ribosomal RNA precursor of Tetrahymena: involvement of a guanosine nucleotide in the excision of the intervening sequence. Cell 27, 487–496.

132 Gilbert, W., 1986. The RNA world. Nature 319, 618.

132 Bernhardt, H.S., 2012. The RNA world hypothesis: the worst theory of the early evolution of life (except for all the others). Biol. Direct. 7, 23.

133 Callahan, M.P., et al., 2011. Carbonaceous meteorites contain a wide range of extraterrestrial nucleobases. Proc. Natl. Acad. Sci. USA 108, 13995–13998.

133 D'Argenio, B., et al., 2001. Microbes in rocks and meteorites: a new form of life unaffected by time, temperature, pressure. Rendiconti Lincei 12, 51–68.

133 Crick, F.H., Orgel, L.E., 1973. Directed panspermia. Icarus 19, 341–346.

133 Parfreya, L.W., et al., 2011. Estimating the timing of early eukaryotic diversification with multigene molecular clocks. Proc. Natl. Acad. Sci. USA 108, 13624–13629.
134 Pace, N.R., et al., 2012. Phylogeny and beyond: scientific, historical, and conceptual significance of the first tree of life. Proc. Natl. Acad. Sci. USA 109, 1011–1018.
134 Domazet-Loso, T., Tautz, D., 2008. An ancient evolutionary origin of genes associated with human genetic diseases. Mol. Biol. Evol. 25, 2699–2707.
134 Sagan, L., 1967. On the origin of mitosing cells. J. Theor. Biol. 14, 225–274.
134 One of Lynn Margulis's sons from her first marriage, Dorion Sagan, became a popular science writer and her collaborator.
134 Rose, S., 2011. Lynn Margulis obituary. The Guardian (Guardian News and Media Limited).
135 Andersson, S.G.E., et al., 1998. The genome sequence of *Rickettsia prowazekii* and the origin of mitochondria. Nature 396, 133–140.
135 Martin, W., et al., 2002. Evolutionary analysis of Arabidopsis, cyanobacterial, and chloroplast genomes reveals plastid phylogeny and thousands of cyanobacterial genes in the nucleus. Proc. Natl. Acad. Sci. USA 99, 12246–12251.
136 Forterre, P., 2013. The virocell concept and environmental microbiology. ISME J. 7, 233–236.
136 Lwoff, A., 1967. Principles of classification and nomenclature of viruses. Nature 215, 13–14.
136 Wessner, D.R., 2010. The origins of viruses. Nat. Edu. 3 (9), 37.

Chapter 15

Evolution: From Darwin to the Hologenome Concept

It is not the strongest of the species that survives, nor the most intelligent, but the one most responsive to change.

—Paraphrase of Charles Darwin in the writings of Leon C. Megginson

Nothing in biology makes sense except in the light of evolution.

—Theodosius Dobzhansky, evolutionary biologist

So, like it or not, microbiology is going to be in the center of evolutionary study in the future—and vice versa.

—Carl R. Woese, an American microbiologist and biophysicist, who is famous for developing biological classification based on 16S ribosomal RNA genes

The word evolution, derived from the classical Latin term *evolution* for the unrolling of a scroll, refers to the transformation of an earlier form or arrangement of something into a different form or arrangement—the present unrolls from the past; the future unrolls from the present. So evolution is about things that change over time. Our subject is the evolution of life—the unrolling of changes in living creatures.

Charles Darwin's theory of evolution by natural selection among variants in the population, also referred to as "survival of the fittest," is the most important idea in biology and has significant impact in other fields, such as sociology, economics, politics, and religion. What was it about Darwin as a person and the circumstances in which he lived that allowed this great idea to emerge in the middle of the 19th century?

Darwin was born in 1809 in Shrewsbury, England to intellectual and wealthy parents, who supported him in his studies and research throughout his life. He received a broad education that included 2 years of Medical School at the University of Edinburgh. Charles found most lectures dull and surgery distressing, so neglected his studies. For stimulation he joined the Plinian Society, a student natural history group, and assisted the comparative anatomist Robert Edmond Grant in his investigations of the physiology and life cycle of marine invertebrates. Darwin presented at the Plinian Society his own discovery that black spores found in oyster shells were the eggs of a leech. He also learned to classify

It's in Your DNA. http://dx.doi.org/10.1016/B978-0-12-812502-1.00015-9

plants and assisted with work on the collections of the University Museum, one of the largest museums in Europe at the time.

Darwin's neglect of his medical studies annoyed his father, who sent him to Christ's College, Cambridge, for a Bachelor of Arts degree as the first step toward becoming an Anglican parson. At Cambridge, Charles became interested in beetle collecting; he pursued this zealously, getting some of his observations published in James Stevens' *Illustrations of British Entomology.* He became a close friend and follower of a botanist, Professor John Henslow, and met other leading naturalists who saw scientific work as religious natural theology. When his own exams drew near, Charles focused on his studies and in his final examination in 1831 he did well, coming 10th out of 178 candidates for the ordinary degree.

After his degree, Darwin studied William Paley's book *Natural Theology or Evidences of the Existence and Attributes of the Deity,* which made an argument for divine design in nature, explaining adaptation as God acting through laws of nature. He also read John Herschel's book, which described the highest aim of natural philosophy as understanding natural laws through inductive reasoning based on observation. By the time Darwin completed his formal education at the age of 22, it was clear that he had a great curiosity of the natural world, an ability to pursue research independently, and a broad education with special interests in biology, geology, and the philosophy of religion.

On returning home from Cambridge University, Darwin found a letter waiting for him from his mentor Professor Henslow, proposing him as a suitable gentleman naturalist for a self-funded place on the ship H.M.S. Beagle with Captain Robert FitzRoy, more as a companion than a mere collector. The ship was to leave in 4 weeks on an expedition to chart the coastline of South America. The voyage, which began on December 27, 1831 and lasted almost 5 years, was to become a major inspiration for Darwin's theory of evolution.

During the expedition, Darwin spent much of his time on land observing animals, plants, and geological formations. He kept careful notes and drawings of his observations and collected many samples. At intervals during the voyage his specimens were sent to Cambridge University for storage together with letters to his family. Darwin had some expertise in geology, beetle collecting, and dissecting marine invertebrates, but in all other areas he was a novice and collected specimens for subsequent appraisal by experts after he returned to England.

On their first stop ashore at St. Jago (now called Santiago; it is the largest island of Cape Verde), Darwin found that a white band high in the volcanic rock cliffs included seashells. Darwin experienced an earthquake in Chile and saw signs that the land had just been raised, including mussel-beds stranded above high tide. High in the Andes he saw seashells and several fossil trees that had once grown on a sand beach. He theorized that as the land rose, oceanic islands sank, and coral reefs around them grew to form atolls (ring-shaped chains of islands formed of coral). Captain FitzRoy had given him the first volume of

Charles Lyell's *Principles of Geology* which set out uniformitarian concepts of land slowly rising or falling over immense periods, and Darwin's own observations strengthened his acceptance of Lyell's theories. Lyell's book clearly influenced Darwin to think of evolution as a slow process in which small changes gradually accumulate over immense spans of time.

After surveying the coasts of South America, the ship stopped over in the Galápagos Islands. The Galápagos Islands are an archipelago of volcanic islands distributed on either side of the Equator in the Pacific Ocean, 906 km (563 miles) west of continental Ecuador, of which they are a part. Darwin noted that the unique creatures were similar, but not identical, from island to island, but perfectly adapted to their environments which led him to ponder the origin of the islands' inhabitants. Among those that struck Darwin were the finches that are now named in his honor. Darwin would later write, "Seeing this gradation and diversity of structure in one small, intimately related group of birds, one might really fancy that from an original paucity of birds in this archipelago, one species had been taken and modified for different ends."

Before describing Darwin's formulation of evolution, I would like to mention the general phenomenon of young people experiencing a different part of the world than they are accustomed to. Moving out of one's environment allows a person to see the world with a fresh eye. As the French novelist Marcel Proust wrote:

> *The real voyage of discovery consists not in seeking new landscapes, but in having new eyes.*

In modern days it is common for young scientists, after completing their PhD degree, to carry out research in a different country, not only to be exposed to new scientific ideas and techniques, but also to open their eyes to a different culture.

On returning to England in 1836, Darwin visited Cambridge to see Professor Henslow, who advised him to find naturalists to help catalog the collections and agreed to take on the botanical specimens himself. Darwin's father organized investments, enabling his son to be a self-funded gentleman scientist, and an excited Darwin went from one British institution to another, seeking experts to more fully describe his collections. While rewriting the notes from his voyage, Darwin edited and published, between 1838 and 1843, a 5-part book, *Zoology of the Voyage of H.M.S. Beagle*, which included reports on his collections by numerous experts on different parts of the collection. Darwin contributed geological introductions to parts of the book.

Continuing his research in London, Darwin read Malthus's *An Essay on the Principle of Population*, which argues that human population, when unchecked, goes on doubling itself every 25 years, a geometric progression so that population soon exceeds food supply in what is known as a Malthusian catastrophe. Darwin compared this to the "warring of the species" of animals and plants and the struggle for existence among wildlife. He reasoned that although the numbers of a species kept roughly stable, species always breed beyond available resources. Thus, favorable variations would make organisms better at surviving and passing the variations on to their offspring, while unfavorable variations would be lost. This would result in the formation of new species. As he later wrote in his autobiography:

> *I happened to read for amusement Malthus on Population, and being well prepared to appreciate the struggle for existence which everywhere goes on from long-continued observation of the habits of animals and plants, it at once struck me that under these circumstances favourable variations would tend to be preserved, and unfavourable ones to be destroyed. The result of this would be the formation of new species. Here, then, I had at last got a theory by which to work.*

In 1858, after years of further scientific investigation, Darwin publically introduced his revolutionary theory of evolution in a letter read at a meeting of the Linnean Society of London. On November 22, 1859, he published a detailed explanation of his theory in his best-known work, *On the Origin of Species by Means of Natural Selection.*

Prior to Darwin, naturalists believed that all species either came into being at the start of the world, or were created over the course of natural history. In either case, the species were believed to remain much the same throughout time. Darwin, however, noticed similarities among species all over the globe, along with variations based on specific locations, leading him to posit that they had gradually evolved from common ancestors. He argued that species survived through a process called "natural selection," where species that successfully adapted to meet the changing requirements of their natural habitat thrived, while those that failed to evolve and reproduce died off. Darwin called his theory natural selection because he was well-aware of "artificial selection" used by farmers to select the best stock in breeding (Fig. 15.1).

Darwin's book *On the Origin of Species* proved unexpectedly popular, with the entire stock of 1250 copies oversubscribed when it went on sale to booksellers and aroused international interest. Although Darwin had only referred to humans in the book by writing "Light will be thrown on the origin of man," the first review claimed it made a creed of the "men from monkeys." In support of Darwin, Thomas Huxley, a famous zoologist often referred to as "Darwin's Bulldog," said that "he would rather be descended from an ape than a man who misused his gifts." Richard Owen, one of the leaders of the scientific establishment at the time, attacked Darwin and his friends, condescendingly dismissing his ideas and began to promote ideas of supernaturally guided evolution. Heated arguments over Darwin's theory of evolution have continued until today. Darwin died of heart failure in 1882 at the age of 73.

By the end of the 19th century most scientists accepted Darwin's revolutionary idea of evolution by natural selection. However, strong opposition to Darwin's theory remains, especially with religious fundamentalists. Opponents of evolution have tried to eliminate the teaching of Darwin's theory from public

FIGURE 15.1 Portrait of Charles Darwin. *(From Origins, Richard Leakey and Roger Lewin. Taken from https://en.wikipedia.org/wiki/Charles_Darwin#/media/File:Charles_Darwin_by_G._Richmond.png.)*

school science curricula or urged science instructors also to teach a version of the creation story found in the biblical book of Genesis. The famous 1925 Scopes "monkey" trial, for instance, involved a Tennessee law prohibiting the teaching of evolution in the state's schools. But beginning in the 1960s, the US Supreme Court issued a number of decisions that imposed severe restrictions on those state governments that opposed the teaching of evolution. As a result of these rulings, school boards, legislatures, and government bodies are now barred from prohibiting the teaching of evolution. Teaching creation science, either along with evolutionary theory or in place of it, is also banned, based on the first amendment to the US Constitution, which separates church and state.

Variation is an essential component of Darwin's theory of evolution by natural selection. It is the raw material for evolution. Without genetic variation, evolution cannot occur—no genetic variation = no evolution. To begin with, variation is a readily observable feature of the biological world—no two sisters or brothers are identical (even identical twins are not identical). Plant and animal breeders have for hundreds of years used variation to develop species with desirable traits. Darwin used his knowledge of biological variation to derive the theory of evolution by natural selection. But what is the origin of variation? Darwin's ideas on variation differ significantly from more recent views. Darwin tended to accept the Lamarckian view that variations arose as a result of the conditions of life.

Jean-Baptiste Lamarck (1774–1829), a renowned French botanist, zoologist, and philosopher of science, published in 1809 in his book *Philosophie Zoologique* that environmental forces lead to change in organisms and these changes were then passed on to future generations. Thus, the main features of his theory of acquired characteristics, referred to today as "Lamarckism," are:

1. Use and disuse—individuals lose characteristics they do not use and develop characteristics that are useful.
2. Inheritances of acquired characteristics—individuals transmit acquired characteristics to offspring.

For example, if an individual develops large muscles by exercise, the larger muscles will be transmitted to offspring, or if a fish lives in a totally dark cave, it loses its visionary processes, and this trait is passed on to its offspring. The fact that Darwin accepted Lamarckism can be seen from the following quote from *The Origin of Species*:

> *I have hitherto sometimes spoken as if the variations so common and multiform in organic beings under domestication, and in a lesser degree in those in a state of nature had been due to chance. This, of course, is a wholly incorrect expression. The much greater variability, as well as the greater frequency of monstrosities, under domestication or cultivation, than under nature, leads me to believe that deviations of structure are in some way due to the nature of the conditions of life, to which the parents and their more remote ancestors have been exposed during several generations.*

Lamarckism was discredited and largely ignored throughout most of the last century. There were two major scientific arguments for rejecting Lamarckism. First, the evolutionary theorist August Weismann argued that inheritance only takes place by means of germ (reproductive) cells and that germ cells cannot be affected by anything other cells of the body acquire during their lifetime. Thus, if you develop muscles or lose a finger by accident, these traits will not be passed on to the next generation.

Second, during the first decade of the 20th century, Hugo de Vries, a Dutch botanist and one of the first geneticists, recognized that a lot of variation in nature is discontinuous, big jumps with no intermediates. This led him to conclude that variation was the result of "mutation," a process that suddenly and without apparent cause irreversibly changed the germ plasma. Mutation produces a biological variant in a single step. In the Modern Synthesis version of Darwinian evolution, beginning in the 1930s, germ plasma became genes, units of heredity, and then genes became DNA sequences. This led to the conclusion that mutations, the ultimate source of variation, are equivalent to changes in DNA sequence, which arise through rare and random errors during DNA replication or are caused by physical or chemical mutagens.

As I will discuss at the end of this chapter, the recently described "Hologenome Concept of Evolution" brings forth novel modes of variation, some of which contain Lamarckian principles within a Darwinian framework.

The debate on units of selection has been a central issue within the field of evolutionary biology from its beginnings. Darwin considered the individual organism as the primary unit of selection in evolution, as did and do many evolutionists till today. However, the codiscoverer of evolution, Alfred R. Wallace, argued that a characteristic can evolve also because it benefits the group even though it may be harmful to the individual possessing it. Subsequently, evolutionary biologists referred to this latter concept as group selection. For many years, group selection, in addition to individual selection, was accepted by evolutionary biologists. Then, George C. Williams published his book *Adaptation and Natural Selection* that claimed biologists had unnecessarily discussed traits existing for the good of the group or species because explanations of the phenomena exist at lower levels. The real unit of selection, according to Williams is neither the group nor the individual, but the gene. Subsequently, Richard Dawkins popularized this concept in his book *The Selfish Gene*.

More recently, a number of prominent evolutionary theorists have come to see the body of ideas known loosely as "multilevel selection theory" as a potent explanatory principle. Multilevel selection theory posits that natural selection can operate simultaneously at different levels of the biological hierarchy. Thus, the individual, the group and the gene can all be units of selection. As we will see in the remainder of this chapter, a new concept of evolution, the hologenome concept, introduces a novel level of selection that expands our understanding of variation and evolution.

Before discussing the hologenome concept, I would like to trace how my partner Ilana Zilber-Rosenberg and I came upon this concept. In 1995, we discovered that the bleaching disease of the coral *Oculina patagonica* in the Mediterranean Sea was caused by a bacterium we named *Vibrio shiloi*. For 7 years we studied how the *Vibrio* caused the disease, publishing a dozen research articles and reviews. Then, to our dismay, the coral became resistant to the *Vibrio*. Our model system was gone.

> *There is no such thing as a failed experiment, only experiments with unexpected outcomes.*
>
> —Richard Buckminster Fuller, American architect and inventor

The unexpected development of resistance of the coral to the *Vibrio* pathogen led us to eventually frame the hologenome concept of evolution. Before giving up and switching to another research topic, we asked the question: how did the coral become resistant? Corals do not have an adaptive immune system and do not produce antibodies. To explain the resistance, we proposed that the coral acquired beneficial bacteria that protected the coral against the *Vibrio* pathogen. My colleagues asked us: how in just a few years is it possible for most of the corals in the Mediterranean Sea to acquire the beneficial bacteria? We replied that if it is possible to have an epidemic of a pathogen, why not an epidemic of a beneficial microbe? It probably happens all the time but goes unnoticed.

One evening while discussing our beneficial bacteria hypothesis over a glass of red wine, it occurred to us that what we had suggested for corals should be true for all animals and plants. This generalization provided the inspiration for the hologenome concept of evolution. To understand the concept, it is necessary first to grasp the meaning of two terms: *holobiont* and *hologenome*. A holobiont, sometimes referred to as a metaorganism, is defined as an animal or plant with all its resident microorganisms. We introduced the term hologenome to describe the total genetic information of a holobiont, the sum of the host genome and all the associated microbial genomes.

The hologenome theory of evolution posits that the holobiont (host + symbionts) with its hologenome (host genome + all the microbial genomes), acting in consortium, function as a unique biological entity and therefore as a level of selection in evolution.

Since the publication of the hologenome theory of evolution in 2008, a large body of empirical data generated by numerous biologists has provided support for the concept. The key points are:

1. All animals and plants harbor abundant and diverse microbiota. In many cases the number of symbiotic microorganisms and their combined genetic information far exceed that of their host. In humans, for example, there are about the same number of microbial cells as human cells and 400 times more unique genes in the diverse symbiotic microbes than in the human genome. Further evidence supporting this point was presented in Chapter 12.
2. The microbiota together with the host genome can be transmitted from one generation to the next with fidelity and thus propagate the unique properties of the holobiont and the species.
3. The microbial symbionts and the host interact in a way that affects the health and fitness of the holobiont within its environment. Several examples of this principle were discussed in Chapter 13, including the fact that symbiotic microbes protect their animal host from infection by pathogens, assist in the digestion of plant fiber and produce essential vitamins and hormones.
4. Genetic variation in the hologenome can be brought about by changes in either the host genome or the microbial population genomes. Under environmental change and stress, the microbial genomes can adjust more rapidly and by more processes than the host organism alone and thus can enhance adaptation and evolution of the holobiont.

Let us now examine how these principles may enhance our understanding of evolution.

If symbionts are to play a role in the evolution of holobionts they must be transmitted from generation-to-generation. Usually, human birth narratives describe the origins of a new individual, focusing on the mother and fetus. The hologenome concept discusses birth as the origin of a new community. Not only is the human genome being transmitted, but so are the genomes of the symbiotic community, whose microbial genes outnumber those of the human host by more than 400-fold. Birth is nothing less than the passage from one set of symbiotic relationships to another.

Symbionts are transferred from parent to offspring by a variety of mechanisms of varying degrees of fidelity. In the case of humans, most of the microbes are acquired from the mother during the birth process when the amnion bursts during labor and the fetus comes into contact with the resident microbes of the birth canal. Subsequently, offspring obtain 10^7–10^8 bacteria daily from their mothers' milk. Additional microbes are obtained from parent kissing and cuddling the baby. For example, in just a single 10-s kiss over 80 million microbes are transferred from the parent to offspring. If the offspring is born by cesarean section (C-section) or is not breast-fed, then kissing and cuddling is particularly important for the offspring to acquire beneficial microbes.

In many different animals, microbiota from the parents is transferred to the offspring by feeding feces to the juveniles. This slightly less direct mode of transmission is used in the termite hindgut-microbiota symbiosis were feces of adult termites (containing abundant microorganisms) are fed to newly hatched juveniles by workers in the colony. Coprophagy is a Greek word meaning feces eating and is used to describe this behavior. In addition to feces-eating insects, many other animals practice coprophagy, including elephants, horses, pandas, koalas, hippos, and rabbits, in order to seed the intestines of their offspring with the bacteria needed to aid in the digestion of local vegetation.

Koalas use a special adaptation of coprophagy; development of the young in the pouch is very slow with the Joey remaining in the pouch for 5–6 months, relying only on its mother's milk. When the Joey is approximately 5 months of age, the mother produces a second type of feces (known as pap) which the Joey eats over several days up to a week. This facilitates introduction of the appropriate gut microbes into the developing juvenile's stomach and cecum and allows for the subsequent digestion of the eucalyptus leaves, which enables eventual weaning from the mother.

Some animals and most plants can reproduce asexually. A striking example is vegetative reproduction in plants. When a fragment of a plant falls to the earth, it may root and grow into a fully developed plant. In such cases, it will clearly contain some of the symbionts of the original plant (direct transfer). In addition, it will most likely incorporate rhizosphere microorganisms from the soil adjacent to the parent.

Regardless of the mechanism used, there is now growing evidence that the microbial component of the holobiont is transferred from generation to generation. The large variations in modes of transmission have an interesting implication: individuals can acquire and transfer symbionts throughout their lives, and not just during their reproductive phase. Furthermore, this implies that the environment (including people you come in contact with) can have an influence on the composition of your hologenome.

Consideration of genetic variation in holobionts leads to the most unique aspect of the hologenome concept. According to the hologenome concept of evolution, genetic variation can arise from changes in either the host or the symbiotic microbiota genomes. In host genomes, variation occurs during sexual reproduction and ultimately by mutation. These same processes occur in microorganisms. However, the microbial component of the hologenome can change by three additional processes: (1) microbial amplification, (2) acquisition of novel strains from the environment, and (3) horizontal gene transfer between different species and between different strains of the same species. These three processes can occur rapidly and are important elements in the evolution of animals and plants.

Microbial amplification is the most rapid and easy to understand mode of variation in holobionts. It involves changes in the relative numbers of the diverse types of associated microorganisms that can occur as a result of nutrient availability, changing temperatures, exposure to antibiotics, and other environmental factors. Increases and decreases of different symbiotic microorganisms lead to changes in the hologenome. Considering the large amount of genetic information encoded in the diverse microbial population of holobionts, microbial amplification is a powerful mechanism for adapting to changing conditions. Numerous studies with humans, nonhuman primates, mice, and other animals have demonstrated that changes in diet or antibiotic treatment can bring about rapid changes in the gut microbiota. Thus, changing your diet or treatment with antibiotics can bring about variation in your hologenome.

Acquisition of novel symbionts from the environment is another way of introducing variation into holobionts. Humans come in contact with billions of microorganisms during their lifetime in the food they eat, water they drink, air they breathe, and by direct interaction with others. It is reasonable to assume that occasionally, as a random event, one of these microorganisms will overcome the immune system, find a niche and become established in the host. Under the appropriate conditions, the novel symbiont may become more abundant and affect the fitness of its new host. Unlike microbial amplification, acquiring new symbionts can introduce entirely new sets of genes into the holobiont.

(When I recently described the role of acquisition of microbes in genetic variation and evolution of animals and plants to a physicist, he remarked: "It makes sense. If I want to improve my iPhone, I don't build an app, but rather install an existing one.")

The "hygiene hypothesis" was originally put forth to explain the increased risk of allergy in both small families and in Western countries. The hypothesis posits that the overly hygienic Western lifestyle limits general microbial exposure and alters the colonization of the infant gut, which in turn disrupts normal development of the immune system and ultimately leads to allergic disease. In support of the hypothesis, it has been shown that reduced diversity of gut microbiota during infancy is associated with allergic disease later in childhood.

The hygiene hypothesis has more recently been expanded to help explain the rise in obesity and related syndromes. Data suggest that improved Westernized sanitation and living conditions, overzealous antimicrobial therapy, delivery by caesarean section, and formula-feeding infants, all of which are widely practiced in developed countries, may predispose individuals to metabolic diseases just as improved hygiene was shown to increase the susceptibility to allergic and autoimmune diseases. In essence, reducing exposure to microorganisms can inhibit acquisition of beneficial symbionts, which have evolved to participate in the metabolism and health of human holobionts (Fig. 15.2).

Horizontal gene transfer (*HGT*) refers to the movement of genetic information across normal mating barriers, between more or less distantly related organisms, and thus stands in distinction to the standard vertical transmission of genes from parent to offspring. HGT is an additional potent mechanism for generating variability in holobionts. Most microbes possess different classes of mobile genetic elements that allow for the acquisition, loss, or rearrangement of sometimes large regions of the hologenome. The evolutionary significance of HGT is that a large block of DNA can be transferred in a single event from one bacterium to a different bacterium or even to the genome of an animal or plant.

An interesting example of variation in the human gut microbiome by HGT is the transfer of a gene coding for the enzyme agarases from a marine bacterium to the human gut bacterium *Bacteroides plebeius* in the Japanese population. The enzyme breaks down the polysaccharide agar that is used to make the nori sheets that wrap around sushi. Agarase is frequent in the Japanese population

FIGURE 15.2 A microbe taking the plunge into a host organism.

but absent in North American individuals. Studies indicate that seaweeds and sushi with associated marine bacteria that contained genes coding for the agarose protein were the source of the enzyme in the Japanese populace. Sometime in the past when they ate raw seaweed, bacteria attached to the seaweed entered their digestive tract, and the agarase-coding gene was horizontally transferred to a resident gut bacterium. Thus, contact with nonsterile food may lead to HGT and variation in human gut microbiota. Although more rare, HGT can also take place from microorganisms to animals and plants and vice versa.

The hologenome concept emphasizes the role of microbiota in the evolution of animals and plants. The hypothesis suggests that animals and plants evolve to a large extent by utilizing the wealth of existing prokaryotic and viral genetic material, in addition to inventing or reinventing genes from scratch. The hologenome concept focuses on cooperation between microbiota and their hosts in the evolution of holobionts, rather than solely on competition between individuals.

Application of modern DNA technology has taught us that animals and plant holobionts contain a wide variety of microbial and viral genetic material, not only in their microbial genes, but also in the host genome. These new DNA techniques have shown us that the collective genomes of human gut bacteria contain 400 times more genes than are present in the human genome. Also, of the approximately 23,000 human genes that code for proteins, 60% arose from bacterial genes. Furthermore, 100,000 pieces of viral DNA have been identified in our genes, making up another 8% of the total human genome. Similar results have been obtained with other animals and plants. Thus, animals and plants have acquired preevolved genetic information and functions from microorganisms, including viruses. Let us now speculate on how these microbial genes were acquired and how complex holobionts may have evolved, and are still evolving.

Higher organisms were and still are constantly exposed to microbes in the environment. A fraction of these microbes attach to the outer surfaces of tissues. If these microbes can find a niche, overcome the immune system and multiply, they become resident symbionts. These symbionts contain whole sets of genes, which become part of the hologenome, and consequently the holobionts which contain these novel genes are variants upon which selection can act, namely, that can be selected for or against. Potentially, newly acquired, useful microbes can either (1) remain intact and express their beneficial genetic potential in the holobiont as part of the microbiota (insourcing), or (2) have some of their genetic material transferred to host chromosomes by horizontal gene transfer (HGT).

An evolutionary important example of microbial insourcing is cellulose degradation in termites and ruminants. In the case of cellulose degradation by termites, it has been suggested that the evolution of a sophisticated community of hindgut microorganisms may be viewed as a gradual process of internalizing mixtures of external anaerobic microbes that digest plant litter in the soil. Instead of plant debris decaying to varying degrees in the external environment prior to ingestion, it "rots" primarily in the hindgut after ingestion. The primary difference between many termites and other invertebrate decomposers is that in termites plant materials (wood, grass, leaf litter) may be consumed before these substrates have been significantly degraded by microorganisms in the soil. Similar arguments have been put forth for the origin of herbivorous dinosaurs and the first plant-eating mammals.

Microbial symbionts not only provide holobionts with an enormous amount of genetic information, but they also have the potential to transfer some of their microbial genes to host chromosomes. There are numerous examples of horizontal gene transfer (HGT) between symbionts and their hosts. One particularly interesting example is the evolution of placental mammals, which of course includes humans. The protein synctin is required for the development of placenta. Synctin is a known viral protein that allows certain viruses to fuse host cells together, enabling spread of the virus from one cell to another. It is believed that placenta mammals were formed by HGT of the gene coding for synctin to a small, insect-eating, scampering animal, shortly after the dinosaurs' demise, 160 million years ago. Now the viral protein allows babies to fuse to their mothers, enabling the tolerance of the fetus by the maternal immune system. Thus, integration of a viral gene into host genomes led to a major evolutionary leap, the formation of placental mammals.

Further support for the hologenome concept of evolution came from an experiment on the origin of species by Brucker and Bordenstein at Vanderbilt University. Species are one of the fundamental units of comparison in virtually all subfields of biology, especially evolution. Ernst Mayr, one of the 20th century's leading evolutionary biologists, articulated the biological species definition in his book *Systematics and the Origin of Species*:

> *Species are groups of interbreeding natural populations that are reproductively isolated from other such groups. The isolating mechanisms by which reproductive isolation is effected are properties of individuals. Hybrids between closely related species are often inviable or, if they live, they are sterile.*

By considering the biological species definition together with the hologenome concept, Brucker and Bordenstein in 2013 reasoned that negative interaction between host genes and host microbiome can accelerate the evolution of hybrid lethality and sterility. They then demonstrated experimentally that microbiota play an important role in wasp speciation. The researchers found that the gut microbiota of two recently diverged wasp species act as a barrier that prevents their evolutionary paths from reuniting. The wasps have significantly different collections of gut microbes and when they cross-breed, the hybrids develop a distorted microbiome that causes their death during the larval stage. Eliminating gut bacteria via antibiotic treatment rescued hybrid survival. Moreover, feeding bacteria to germ-free hybrids reinstated lethality. The authors conclude: "In this animal complex, the gut microbiome and host genome represent a co-adapted hologenome that breaks down during hybridization, promoting hybrid lethality and assisting speciation."

Although in my opinion there is considerable experimental data supporting the hologenome concept of evolution, I must acknowledge that there are some eminent biologists that challenge the concept, primarily because there is insufficient evidence to generalize about the transmission of microbiomes between generations. Clearly, more experiments are necessary to resolve this question. On a personal note, I am encouraged by a quotation from the physicist Max Planck:

> *A scientific truth does not triumph by convincing its opponents and making them see the light, but rather because its opponents eventually die and a new generation grows up that is familiar with it.*

Let me summarize this chapter by indicating that the Darwinian "survival of the fittest" as a synonym for natural selection is a harsh conviction for the biological world when it comes to individuals and even more so for human society. The hologenome concept of evolution suggests a less abrasive mode of evolutionary change, emphasizing cooperation hand in hand with competition. Cooperation has been selected at every level in evolution, from genes to holobionts, groups of holobionts and human societies. The historian Yuval Harari in his book, *From Animals Into Gods: A Brief History of Humankind,* has argued that the ability to cooperate effectively in large groups has made *Homo sapiens* the masters of planet Earth. Thousands of humans, working together, created trade routes, mass celebrations, political institutions, and technology. A modern jet aircraft is produced through the cooperation of thousands of strangers all over the world: from workers who mine the metals to engineers who test the aerodynamics. To ignore cooperation in human behavior, as well as in the rest of the biology world, is to neglect one of the greatest attributes of living organisms. Hopefully, recognition of the hologenome concept of evolution by cooperation will moderate the selfish view of Social Darwinism.

NOTES AND REFERENCES

139 Leon, C., 1963. Megginson wrote According to Darwin's Origin of Species, it is not the most intellectual of the species that survives; it is not the strongest that survives; but the species that survives is the one that is able best to adapt and adjust to the changing environment in which it finds itself. Lessons from Europe for American Business, Southwestern Social Science Quarterly 44, 3–13.

139 Dobzhansky, T., 1973. Nothing in biology makes sense except in the light of evolution. Am. Biol. Teacher 35, 125–129.

139 The English philosopher and biologist Herbert Spencer actually coined the expression "survival of the fittest."

141 Darwin's finches (also known as the Galápagos finches) are a group of about 15 species of birds. All are found only on the Galápagos Islands, except the Cocos finch from Cocos Island. The term "Darwin's finches" was first applied by Percy Lowe in 1936, and popularized in 1947 by David Lack in his book *Darwin's Finches*.

141 Marcel Proust: French novelist, best known for his 3000 page masterpiece Remembrance of Things Past.

142 The Linnean Society of London is the world's oldest biological society. It is a society for the study and dissemination of taxonomy and natural history.

144 Weismann, A., 1893. The Germ-Plasm: A Theory of Heredity. Charles Scribner's Sons. Electronic Scholarly Publishing, New York, NY, USA.

144 Stamhuis, I.H., et al., 1999. Hugo de Vries on heredity, 1889–1903. Statistics, Mendelian laws, pangenes, mutations. Isis 90, 238–267.

145 Ruse, M., 1980. Charles Darwin and group selection. Ann. Sci. 37, 615–630.

145 Okasha, S., 2006. The levels of selection debate: philosophical issues. Philos. Compass 1, 1–12.

145 Muir, W.M., et al., 2013. Multilevel selection with kin and non-kin groups. Experimental results with Japanese quail (*Coturnix japonica*). Evolution 67, 1598–1606.

146 Zilber-Rosenberg, I., Rosenberg, E., 2008. Role of microorganisms in the evolution of animals and plants: the hologenome theory of evolution. FEMS Microbiol. Rev. 32, 723–735.

147 Gilbert, F.S., 2014. A holobiont birth narrative: the epigenetic transmission of the human microbiome. Front. Genet. 5, 282.

Musso, G., et al., 2010. Obesity, diabetes, and gut microbiota: the hygiene hypothesis expanded? Diabetes Care 33, 2277–2284.

149 Hehemann, J.H., et al., 2010. Transfer of carbohydrate-active enzymes from marine bacteria to Japanese gut microbiota. Nature 464, 908–914.

151 Dupressoir, A., et al., 2012. From ancestral infectious retroviruses to bona fide cellular genes: role of the captured syncytins in placentation. Placenta 33, 663–671.

151 Brucker, R.M., Bordenstein, S.R., 2012. Speciation by symbiosis. Trends Ecol. Evolut. 27, 443–451.

152 Wissenschaftliche Selbstbiographie. Mit einem Bildnis und der von Max von Laue gehaltenen Traueransprache. Johann Ambrosius Barth Verlag (Leipzig 1948), p. 22, as translated in Scientific Autobiography and Other Papers, Gaynor, F. (Trans.) (New York, 1949), pp. 33–34 (as cited in T. S. Kuhn, The Structure of Scientific Revolutions).

Chapter 16

DNA and Cancer

Cure sometimes, treat often, comfort always.

—Hippocrates (460–370 BC)

Understanding how DNA transmits all it knows about cancer, physics, dreaming and love will keep man searching for some time.

—David R. Browe, prominent environmentalist and the first Director of the Sierra Club

Your genome knows much more about your medical history than you do.

—Danny Hillis, American inventor, scientist, engineer, and author

Cancer is the general name for a group of many diseases. Although there are many kinds of cancer, all cancers are characterized by abnormal cells that divide out of control. Cancer cell growth is different from normal cell growth. Instead of dying, cancer cells continue to divide and form new, abnormal cells. Cancer cells can also grow into other tissues, something that normal cells can't do. Multiplying out of control and invading other tissues are what makes a cell a cancer cell.

The oldest description of cancer in humans was found in an Egyptian papyrus written in 1600 BC. It referred to eight "ulcers" (cancers) of the breast. Hippocrates—the "Father of Medicine" who lived in Greece around 400 BC—is credited with being the first to recognize the difference between benign and malignant tumors. In fact, the name cancer comes from the ancient Greek word for crab, as scientists at the time thought that clusters of cancer cells looked like the legs of a crab.

Before discussing the connection between DNA and cancer, it is worthwhile to consider some of the environmental causes of cancer. The first to suggest a cause for one type of cancer was the English surgeon Sir Percivall Pott, who in 1775 noticed that chimney sweeps in London at the time were boys, some as young as 4 years old, and many of them would get cancer of the scrotum (part of the external male genitalia located behind and underneath the penis) in their late teens or early 20s. He suggested that something in the soot was causing cancer. By pointing out an association between exposure to soot by chimney sweeps and cancer of the scrotum, Percivall Pott contributed to the creation of the field

It's in Your DNA. http://dx.doi.org/10.1016/B978-0-12-812502-1.00016-0

of epidemiology, the branch of medicine dealing with the incidence and prevalence of disease in populations and with the detection of possible sources and causes of disease epidemics. As I shall show, most of the important causes of cancer were first recognized from epidemiological studies.

It should be pointed out that showing a correlation between exposure to coal tar and cancer does not demonstrate cause-and-effect. For example, there is a strong correlation between ice cream sales and the rate of drowning deaths. But increase in ice cream consumption does not cause drowning. Ice cream is sold during the hot summer months at a much greater rate than during colder times, and it is during these hot summer months that people are more likely to go swimming. The increased drowning deaths are simply caused by more exposure to water-based activities not ice cream. A cause-and-effect relationship between soot and scrotal cancer was first indicated indirectly. In 1788, the Chimney Sweepers Act was passed in the British Parliament, preventing the employment of children under eight. Furthermore, Pott put forth the revolutionary recommendation that people should bathe once a day. By 1875, the use of young chimney sweeps was completely forbidden. As a result, the epidemic of scrotal cancer in seeps gradually ceased.

Pott's hypothesis that exposure to coal tar was the cause of cancer in chimney sweeps was validated directly many years later when, in 1915, Japanese scientists reported that after they had applied extracts of coal tar to the skin of 137 rabbits every 2 or 3 days for 3 months, 7 of the rabbits developed cancers at the application sites. Eventually it was shown that several hydrocarbons (chemicals containing only hydrogen and carbon) present in coal tar caused cancer in animals. The most active cancer-causing chemical in coal tar was benzopyrene. A substance or agent that tends to produce cancer is referred to as a carcinogen. The carcinogen benzopyrene is a hydrocarbon consisting of five fused rings (Fig. 16.1).

Sixty-five years ago cancer was considered to be a natural consequence of old age, much like gray hair, wrinkled skin, hearing loss, and osteoporosis. Then, in 1950, Ernst Wynder and Evarts Graham in the United States and Richard Doll and Bradford Hill in England published at about the same time epidemiological studies that indicated that smoking was a cause, and an important cause, of the rapidly increasing epidemic of lung cancer. Comparing 649 lung cancer patients with 600 other patients (controls), Wynder and Graham found lung cancer an incredible 40 times higher among smokers, with the risk of cancer increasing

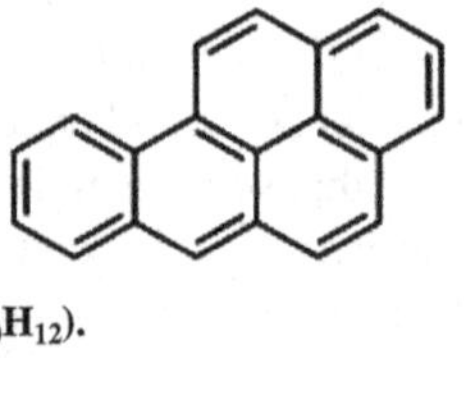

FIGURE 16.1 Benzopyrene ($C_{20}H_{12}$).

with the number of cigarettes smoked. Doll and Hill later confirmed these results using 1465 cases.

In spite of these studies, most of the medical and scientific community in the early 1950s, many of whom themselves were smokers, still did not accept that smoking could cause lung cancer and argued that there must be other explanations for the association. Doll and Hill understood, of course, that a few nonsmokers would get lung cancer (so smoking was not a necessary cause of the disease) and that many smokers would not get cancer (so smoking was not a sufficient cause of the disease). However, that among otherwise similar individuals, smokers had a substantial higher probability of developing cancer, suggesting that smoking was an important cause of the disease. Although Doll and Hill understood that there was already strong circumstantial evidence that smoking was an important cause of lung cancer, they also understood that further research was needed, partly to help convince sceptics and partly to see whether smoking also caused other diseases.

In 1954, Doll and Hill reported an even more convincing finding linking smoking to lung cancer. They sent out questionnaires to British physicians in 1951 to collect details of their smoking habits, and received 25,000 replies from male doctors over the age of 35. They then determined how many of these doctors died during the following 3 years and the causes of their deaths. There were 789 deaths, 36 of which were due to lung cancer. All 36 were smokers, 25 cigarette smokers, 4 pipe smokers, and 7 that smoked both cigarettes and pipes. The results were so clear that Doll and Hill noted that you don't even need statistics to prove the point. There was a much higher mortality from lung cancer in smokers than in nonsmokers, and a clear dose–response relationship between the amount smoked and the death rate from lung cancer. Subsequent results from the doctor study indicated a progressive and significant reduction in mortality with the increase in the length of time over which smoking had been given up. There was remarkably little difference between the smoking habits of doctors who lived in large towns and those who lived in other districts, so Doll and Hill concluded that lung cancer could not be attributed to a differential exposure to atmospheric pollution.

To combat the claim that smoking is bad for your health, the tobacco industry spent hundreds of millions of dollars advertising cigarettes and trying to convince the public and government authorities that smoking was neither the cause of cancer nor was addictive. Two typical advertisements aimed at assuring smokers of the safety of smoking were: "More doctors smoke Camels" and "Play safe with Philip Morris." Other catchy slogans are still used today: "Winston tastes good like a cigarette should!" and "Us Tareyton smokers would rather fight than switch!" By 1965, the fraction of smokers in the adult American population reached an all-time high of 45%, and the gross annuals sales of cigarettes in the United States was $5 billion.

As the evidence linking smoking to cancer continued to increase, governments slowly began to introduce laws aimed at reducing smoking. In 1964, the

"Smoking and Health: Report of the Advisory Committee to the Surgeon General of the United States" was published. It was based on over 7000 scientific articles that linked tobacco use with cancer and other diseases. This report led directly to laws prohibiting tobacco advertisements on television and subsequently to requiring warning labels on tobacco products. As these came into force and there was a growing awareness of the dangers of tobacco, smoking in America decreased slowly but significantly. By 2014, half of all Americans who had ever smoked had quit. Nevertheless, 160,000 American smokers still died of lung cancer in 2014 and 220,000 new cases were reported according to the American Cancer Society. Worldwide, lung cancer kills over one million people each year. Lung cancer is the leading cancer killer in both men and women in the United States. Lung cancer causes more deaths than the next three most common cancers combined (colon, breast, and pancreatic). More than four in five lung cancers are caused by smoking.

In 1996, a study was published that provided the clear molecular evidence conclusively linking components in tobacco smoke to lung cancer. Benzopyrene, the same compound that was previously shown to be responsible for cancer in coal tar, was found in tobacco smoke. Smoking a packet of 20 cigarettes results in the inhalation of 400 ng (1 billion ng = 1 g) of the carcinogen benzopyrene. Tiny though this figure may seem it is sufficient over time to exert a damaging effect. In addition to benzopyrene, more than 50 other chemicals in cigarettes have been shown to cause cancer in animals.

Radiation is another important cause of cancer. The types of high energy radiation that are known to cause cancer are ultraviolet light, X-rays, gamma rays, and cosmic rays. Low energy radio frequency radiations from mobile phones, electric power transmission, and other similar sources have been suggested as possible carcinogens by the World Health Organization, but the links remains unproven.

The first source of radiation linked to cancer was X-rays. The German physicist Wilhelm Röntgen was the first to detect and produce X-rays, an achievement that earned him the first Nobel Prize in Physics in 1901. Many scientists and physicians began to experiment with X-rays and it was quickly found that the rays could damage biological tissue and cause leukemia and highly malignant tumors on exposed skin. A total of 360 of these pioneer radiation workers from 22 countries, subsequently referred to as radiation martyrs, died from the cancer. One of these pioneers was Thomas Edison's assistant, Clarence Dally, who helped develop the X-ray fluoroscope. After submitting to amputation of his burned and ulcerated arms, he died from cancer in 1904.

During the first half of the 20th century, X-rays were used by enthusiasts in applications ranging from treatment of acne to removal of unwanted facial hair. This enthusiasm persisted despite the scientific community's early knowledge of the adverse effects of X-ray overexposure. Mobile X-ray machines were used extensively in the field during World War I (1914–18) to locate shrapnel and help set broken bones.

Today, 7 out of 10 people receive either a medical or dental X-ray every year making X-rays the primary source of man-made radiation exposure. Clearly, X-rays are a useful diagnostic tool because they can provide a clear picture of many internal structures that could only be viewed through more complex or intrusive methods without X-rays. Because they are quick and easily performed tests, X-rays are good for situations, such as emergency room visits, where speed is important. However, X-rays have the potential to cause cellular damage that could lead to cancer. X-rays are particularly dangerous for pregnant women and very small children.

It should be pointed out that a single CT (computed tomography) scan subjects the human body to between 200 and 1000 times the X-radiation of a conventional X-ray, or around a year's worth of exposure to radiation from both natural and artificial sources in the environment. Nevertheless, the general consensus among scientists and physicians is that in most cases, the benefits of diagnosis outweigh the possible risks of using X-rays.

The element radon is responsible for the majority of the natural exposure of the public to cancer-inducing radiation. It is often the single largest contributor to an individual's background radiation dose. Radon is a chemical element with symbol Rn and atomic number 86. It is a radioactive, colorless, odorless, tasteless, gas, occurring naturally as a decay product of uranium. Since uranium is essentially ubiquitous in the earth's crust, radon is present in almost all rocks, soils, and water. However, the concentration of radon is variable from location to location. Radon gas from natural sources can accumulate in buildings, especially in confined areas, such as attics, and basements. It can also be found in some spring waters and hot springs.

Epidemiological evidence shows a clear link between lung cancer and high concentrations of radon, with 21,000 radon-induced US lung cancer deaths per year—second only to cigarette smoking. Thus in geographic areas where radon is present in heightened concentrations, radon is considered a significant indoor air contaminant. All the evidence supports the conclusion that residential exposure to radon can increases the risk of lung cancer in a dose–response manner, that is, the higher the concentration of radon, the greater the chance of causing lung cancer.

The survivors of the atomic bombings in Hiroshima and Nagasaki are a general population of all ages and sexes and, because of the wide and well-characterized range of doses received, have been used as the basis of population cancer risk estimates following radiation exposure. Leukemia (cancer of the body's blood-forming tissues) was the first cancer to be associated with atomic bomb radiation exposure. An excess of solid cancers became apparent approximately 10 years after radiation exposure. For most solid cancer sites a linear dose response was observed. Risks among occupationally exposed groups, such as nuclear workforces and underground miners are generally consistent with those observed in the Japanese atomic bomb survivors.

Most skin cancers, including melanoma, are a direct result of exposure to the ultraviolet (UV) radiation from sunlight. The incidence rate is higher in fairer skinned, sun-sensitive rather than darker skinned less sun-sensitive people. The most common types of skin cancer tend to be found on sun-exposed parts of the body, and their occurrence is related to lifetime sun exposure. Skin cancer has also been linked to exposure to some artificial sources of UV, such as from tanning machines. Based on animal and human evidence the World Health Organization has classified UV radiation as a carcinogen.

The connection between cancer and DNA became evident when it was recognized that all agents that cause cancer (carcinogens) also cause a change in the DNA sequence and thus are mutagens. All the effects of carcinogenic chemicals and radiation on tumor production can be accounted for by the DNA damage that they cause and by the errors introduced into DNA during the cells' efforts to repair this damage. Cells become cancer cells because of DNA damage. In a normal cell, when DNA is damaged the cell either repairs the damage or dies. In cancer cells, the damaged DNA is not repaired, but the cell doesn't die like it should. Instead, the cell goes on making new cells that the body doesn't need. These new cells all have the same damaged DNA as the first abnormal cell does, at least during the early stages of cancer.

The realization that all chemicals that cause mutations also cause cancer led the microbiologist Bruce Ames at the University of California at Berkeley to develop a simple and inexpensive test for determining if any particular substance is carcinogenic. The test used the bacterium *Salmonella typhimurium.* When *Salmonella* is isolated from nature, the wild type strain can grow on a minimal medium that contains only a sugar and salts. Ames began by isolating mutants of *Salmonella* that required the amino acid histidine for growth. The technique that he used to isolate these mutant bacteria was similar to the method that Beadle and Tatum had used to obtain mutants of the mold *Neurospora* (described in Chapter 6). The histidine-requiring *Salmonella* bacteria are referred to as tester strains. The property of these tester strains is that they will grow and produce visible colonies in one day at 37°C on a solid agar medium that contains a sugar, salts, and histidine, but will not form colonies if the histidine is omitted. For a tester strain to form a colony on the minimal agar medium without histidine there must be a second mutation that reverses the effect of the first mutation. Thus, when a tester strain is placed on a minimal agar medium the numbers of colonies that appear reflect the number of back mutations. For example, when one-billion histidine-requiring mutants were placed on the agar minimal medium, approximately five colonies appeared, corresponding to a spontaneous mutation frequency of 5×10^{-9}. However, if the tester strain was exposed to UV light for 1 min prior to placing it on the minimal medium, then 500 colonies appeared, indicating that the radiation caused the frequency of mutations to increase 100 times. Thus UV light is a potent mutagenic and carcinogenic agent.

There was one serious problem with the initial Ames test. Some chemicals, such as benzopyrene, did not cause mutations in the tester strains, even though

FIGURE 16.2 **Bruce N. Ames (1928–).** *(Taken from https://en.wikipedia.org/wiki/Bruce_Ames#/media/File:Bruce_Ames.jpg)*

they were known to cause cancer. Ames reasoned that animals contain certain enzymes that are absent in bacteria and these enzymes could convert benzopyrene into a chemical that was a carcinogen and a mutagen. To test this hypothesis, he added a rat liver extract, which contains a mixture of animal enzymes, to the benzopyrene before applying it to the tester strain and then placing the mixture on the minimal agar medium. This time there were hundreds of colonies, indicating that benzopyrene was, in fact, a mutagen but only after it came into contact with enzymes present in animals. The Ames test and the subsequent modifications of the test now include (1) the *Salmonella* histidine-requiring tester strain, (2) rat-liver extract, and (3) the substance that is being examined for its ability to increase the frequency of mutation (Fig. 16.2).

The simplicity, sensitivity, and accuracy of the Ames test have resulted in its being adopted in hundreds of laboratories around the world, in order to test for environmental mutagens. Essentially all of the chemicals that give positive Ames tests also induce cancer in mice. This is of enormous practical significance since it is inordinately expensive and time-consuming to screen a large number of materials for cancer-producing potentials in animals. To test a single material for its ability to cause cancer in animals requires 2–3 years, costs about $200,000, and necessitates sacrificing several hundred laboratory animals. With the development of Ames test, thousands of environmental materials can be screened for potential carcinogens rapidly, inexpensively, and without having to kill animals.

In 1975, undergraduates at the University of California, under the supervision of Ames, examined several hundred commercial products for potential

FIGURE 16.3 **Use of bacteria to test chemicals and foods for mutagenicity according to the Ames test.**

carcinogens. They found that almost all the hair dyes sold on the US market at that time contained highly mutagenic substances. As a result of that laboratory student exercise, the cosmetics industry was forced to remove the dyes from the market and replace them with new and safer formulations. Today, an increasing number of industries are utilizing the Ames test to examine their products (Fig. 16.3).

The strong connection between cancer and mutation leads directly to the question: how do mutations in DNA lead to cancer? To begin with, it is now clear that not all mutations result in cancer. Genes, which when mutated, contribute to cancer are referred to as *proto-oncogenes* prior to the mutation, and as *oncogenes* after they have been mutated. The initial event in the development of cancer is the mutation of a normal proto-oncogene to produce an oncogene. This mutation is usually due to a change in the base sequence of the proto-oncogene, but it also can occur by the proto-oncogene "jumping" to a different place on the chromosome. The fact that genes can move from one place on the chromosome to another was first shown in corn by Barbara McClintock, as discussed in Chapter 9. Unrepaired mutations build up over a person's lifetime and eventually lead to uncontrolled cell multiplication, that is, cancer. Some of the proto-oncogenes normally function to prevent cell growth, and when they are mutated to oncogenes, control over cell growth is lost, and cancer develops.

In the late 1970s, J. Michael Bishop and Harold Elliot Varmus isolated the first human oncogene, referred to as SRC, and showed that cancer-promoting genes are not foreign intruders, but important normal cell genes that can wreak havoc when mutated. At the time of their discovery, Varmus was carrying out postdoctoral studies in Bishop's laboratory at the University of California, San Francisco. Varmus had an unusual background for a scientist. His BA was in English literature from Amherst College and his graduate degree was in English from Harvard University. After being rejected twice by the Harvard Medical School, he applied and was accepted at Columbia University College of Physicians and Surgeons. On completion of his MD, Varmus worked at a missionary hospital in Bareilly, India. Three years later, in 1970, he began research in Bishop's laboratory.

Bishop and Varmus demonstrated that the gene SRC was, in fact, an oncogene because introduction of the gene into chickens (via a virus) caused the development of tumors. Furthermore, the protein product of the human SRC gene, referred to as c-Src, was found in abnormally high concentrations in a wide variety of human cancers. c-Src is an enzyme that modifies certain proteins by placing a phosphate group on the amino acid tyrosine. The result of the high concentration of the enzyme c-Src is a 10-fold elevation in proteins containing phosphorous in cancer cells compared to those produced by normal cells. The addition of phosphorus to certain proteins affects the control of cell growth and presumably leads to unregulated growth and cancer. Increased activity of c-Src has been shown to increase growth rates and invasion characteristics of tumors, and to decrease death in tumor cells. This has led scientists to begin testing molecules that inhibit the enzymatic activity of c-Src in order to retard tumor growth and reduce the spread of the cancer. In 1989, Bishop and Varmus received the Nobel Prize in Physiology or Medicine for their discovery of the cellular origin of oncogenes.

Since the 1980s, hundreds of oncogenes have been identified in human cancer. So far, analyses of genomic alterations in multiple tumor types have led to three fundamental observations: (1) the mutation that causes a proto-oncogene to become an oncogene is usually a single DNA base substitution. Following the formation of the oncogene there is a latent period of months to years before the cancer becomes evident, suggesting that alterations in other genes must occur before progression to cancer can occur—activation of a particular oncogene seems to be necessary but not sufficient for the development of overt cancer; (2) tumors originating in the same organ or tissue vary substantially in genomic alterations; (3) similar patterns of genomic alteration are observed in tumors from different tissues of origin. These phenomena of intra-cancer heterogeneity and cross-cancer similarity represent both a clinical challenge and an opportunity to design new therapeutic protocols based on the genomic traits of tumors.

What can be done to prevent and treat cancers? Since the initial event in carcinogenesis is the mutation of a proto-oncogene to produce an oncogene, reducing exposure to mutagenic agents, such as radiation and chemical carcinogens, will greatly minimize the probability of getting cancer. However, a certain irreducible background incidence of cancer is to be expected regardless of circumstances: mutations can never be absolutely avoided because they are an inescapable consequence of low frequency errors that occur during DNA replication. If a human could live long enough, it is inevitable that at least one of her or his cells would eventually accumulate a set of mutations sufficient for cancer to develop. Nevertheless, it has been estimated that 80%–90% of cancers could be avoidable, or at least delayable.

The common types of cancer treatment are surgery, chemotherapy, and radiation therapy, either alone or in combination. Five-year survival rates in the United States for pancreatic (5.6%), liver (13.8%), esophagus (16.9%), stomach (25.9%), and brain (26.1%) cancer, remain abysmally low. Even prostate and breast cancers, which are highly amenable to treatment with 5-year survival rates better than 80%, still respond poorly to treatment at later stages. Furthermore, current treatments often have far reaching negative side effects. The systemic toxicity of chemotherapy and radiation regimens, while not as severe as they once were, still often result in acute and delayed nausea and an increased risk of developing other types of cancers. Therefore, entirely new treatment methods are called for in order to alleviate the suffering and death caused by cancers.

Modern methods for detecting and treating cancer are based on knowledge of the DNA sequence of oncogenes. Individuals with a family history of cancer are often counseled to test for mutations in genes known to increase the risk for the disease. In 2014 about 100,000 people underwent these tests. The costs, $1500–$4000, are covered by some, but not all, insurers. Being aware that a particular individual has a proto-oncogene or an oncogene gives that person the option for preventative actions, like getting screened more often, taking a drug to lower the risk, or having their breasts or ovaries removed.

New cancer treatments include targeting the enzymatic activity of oncogene products. Several new drugs, small molecules, and antibodies directly affecting oncogene products have been developed, and more will follow. The advantage of targeted therapy is the dependency of cancer cells on the oncogene product for growth and survival. Thus, cancer cells are more sensitive to the treatment than are normal cells. All targets, however, are not equivalent. It is possible to foresee the development of multiple drugs that have multiple targets involved in the development of cancer.

One example of targeted cancer therapy involves an oncogene, named ERBB2, which is located on the long arm of human chromosome 17. ERBB2 codes for the protein HER2. HER2 is produced in large amounts in approximately 20% of breast cancers and is strongly associated with increased disease recurrence and a poor prognosis. An antibody that specifically attacks HER2 (marketed as Herceptin) has been produced and tested in 13,000 women with early-stage HER2-positive breast cancer. One year treatment with Herceptin resulted in a 25% reduction in disease recurrence after 8 years. A study published in 2014 found that adding a year of Herceptin treatment to standard chemotherapy improved overall survival by 37% but, those benefits don't come without a cost. Herceptin is expensive, with a 1-year course costing approximately $64,000, according to Genentech, the company that makes the drug.

Immunotherapy is the concept of boosting the immune system to target and destroy cancer cells. Tumors grow by subverting the immune system into supporting and protecting the growing tumor from immune surveillance. Therefore the therapeutic aim of immunotherapy is to deprive the growing tumor of its illicit activation of immune suppression and to unleash an autoimmune disease targeted to the tumor. However, limited success has been achieved with traditional immunotherapy, as cancer cells tend to evolve mechanisms that evade immune detection. A wide array of gene therapy techniques are being used to overcome this limitation. Currently, gene therapy is being used to create cancer vaccines. Unlike vaccines for infectious agents, these vaccines are not meant to prevent disease, but to cure or contain it by training the patient's immune system to recognize and destroy the cancer. Initially, cancer cells are harvested from the patient and then are grown in the laboratory. These cells are then genetically engineered to be more recognizable to the immune system by the addition of one or more genes, which produce proinflammatory immune stimulating molecules. These engineered cells are grown in vitro and killed, and the cellular contents are incorporated into a vaccine. These next generation vaccines are already in clinical trials for several cancer types.

Through scientific research, we will eventually find a way to prevent and treat most cancers. This will not necessarily mean inventing a magic bullet to cure all cancers. But it will mean preventing more cases, detecting it earlier, stopping it spreading and—perhaps most crucially of all—finding a way to treat cancer after it's spread. In this regard, the former Director of the National Cancer Institute, Andrew von Eschenbach, wrote: "the cancer cell is only part of the story in cancer development. Mounting evidence now suggests that a cancer cell interacts with its local and systemic microenvironments, and each profoundly affects the behavior of the other." Molecular signals from the microenvironment can influence the expression of cancer genes. Isaac Witz of Tel Aviv University, one of leading proponents of the importance of microenvironments on cancer development, has suggested that targeting the interactions between these signals and cancer cells could reverse the promalignancy effects of the microenvironment.

NOTES AND REFERENCES

155 Dobson, J., 1972. Percivall Pott. Ann. R. Coll. Surg. Engl. 50, 54–65.

157 Wynder, E.L., Graham, E.A., 1950. Tobacco smoking as a possible etiologic factor in bronchiogenic carcinoma; a study of 684 proved cases. J. Am. Med. Assoc. 143, 329–336.

157 Doll, R., Hill, A.B., 1950. Smoking and carcinoma of the lung: preliminary report. Br. Med. J. 2, 739–748.

157 Parkin, D.M., et al., 2011. The fraction of cancer attributable to lifestyle and environmental factors in the UK in 2010. Br. J. Cancer 105, S77–S81.

158 Markel, H., 2007. Tracing the cigarette's path from sexy to deadly. The New York Times.

158 Denissenko, M.F., et al., 1996. Preferential formation of benzo[a]pyrene adducts at lung cancer mutational hotspots in P53. Science 274, 430–432.

Pfeifer, G.P., et al., 2002. Tobacco smoke carcinogens, DNA damage and p53 mutations in smoking-associated cancers. Oncogene 48, 7435–7451.

159 Storrs, C., 2013. How much do CT scans increase the risk of cancer? Researchers reevaluate the safety of radiation used in medical imaging. Sci. Am. 309, 30–32.

160 Ames, B., et al., 1973. Carcinogens are mutagens: a simple test system combining liver homogenates for activation and bacteria for detection. Proc. Natl. Acad. Sci. USA 70, 2281–2285.

161 McCann, J., et al., 1975. Detection of carcinogens as mutagens in the *Salmonella*/microsome test: assay of 300 chemicals. Proc. Natl. Acad. Sci. USA 72, 5135–5139.

162 Adamson, E.D., 1987. Oncogenes in development. Development 99, 449–471.

163 Varmus is currently Director of the National Cancer Institute, a post he was appointed to by President Barack Obama.

164 Klein, A., et al., 2015. Astrocytes facilitate melanoma brain metastasis via secretion of IL 23. J. Pathol. 236, 116–127.

165 Cohen, I.R., 2014. Activation of benign autoimmunity as both tumor and autoimmune disease immunotherapy: a comprehensive review. J. Autoimmun. 54, 112–117.

Chapter 17

DNA, Aging, and Death

Every man desires to live long; but no man would be old.

—Jonathan Swift, *Gulliver's Travels*

It remains for us to discuss youth and age, and life and death. To come to a definite understanding about these matters would complete our course of study on animals.

—Aristotle, *On Longevity and Shortness of Life*

When my daughter was 4, she asked my father how it feels to be so old. His immediate reply was: "Considering the alternative, I prefer it." One of the major goals of medical science is to increase the length and quality of life. Aging can be defined as cellular senescence, which results in a diminished ability to respond to stress, progressive functional decline, greater risk of diseases, such as cancer, atherosclerosis, cardiovascular disease, arthritis, osteoporosis, type 2-diabetes, and Alzheimer's disease, and eventually death. Before discussing some theories on the causation of aging, let us examine some statistics on life expectancy (Table 17.1).

"Life expectancy at birth" is the average number of years that a newborn baby can expect to live in a given society at a given time. The average life expectancy at birth for humans from prehistory to the middle ages was approximately 30 years. It only reached around 40 years in 1850. Since 1850, life expectancy has continuously increased in the Western world, reaching 51, 63, and 81 years in 1900, 1950, and 2014, respectively, for women in the United States of America. In China, the major increase in life expectancy occurred after 1960. There are great variations in life expectancy between different parts of the world, mostly caused by differences in public health, medical care, and diet. Global life expectancy increased by 5 years between the turn of the millennium and 2015, the sharpest increase since the 1960s. There remains significant disparity between wealth nations and poorer ones, with 29 high-income states averaging over 80 years and 22 sub-Saharan African countries averaging less than 60. Today, life expectancy is longest for Japanese women (85.8 years) and shortest for Sierra Leone men (46.0 years). In all countries, women, on average, live 2–6 years longer than men.

It's in Your DNA. http://dx.doi.org/10.1016/B978-0-12-812502-1.00017-2

TABLE 17.1 Life Expectancies at Birth[a]

Calendar period	Region	Women	Men
Bronze age (3200–600 BCE)	Europe		26
Classical Greek (300–500 BCE)	Greece		28
Medieval (400–1500)	Europe		30
1850	USA	40.5	38.3
1900	USA	51.1	48.2
1950	USA	62.7	56.3
2000	USA	80.0	74.8
2014	USA	80.9	76.1
1960	China	45.1	41.9
2014	China	76.8	74.2
2014	Japan	85.8	79.5
2014	Switzerland	84.7	80.3
2014	France	85.0	78.5
2014	Israel	84.3	80.6
2014	India	67.3	63.8
2014	Kenya	64.1	58.1
2014	Sierra Leone	47.0	46.0

[a]*Data from Encyclopedia of Population (2003) and the World Health Organization.*

It would appear that improved conditions during the 20th century in the West and in the last 50 years in China and some other countries are largely responsible for increasing life expectancies. The most important factors that were responsible for raising life expectancies during the last century were reducing infant and mother mortality rates during birth and childhood diseases. For example, the percentage of children that died before the age of 15 in Europe in 1850 was 21.7%, whereas in 2000 it was less than 1%. Other important factors that led to increased life expectancies were availability of vitamins and better nutrition, vaccination procedures, water purification, blood transfusions, antibiotics, and surgery.

It should be pointed out that life expectancy should be distinguished from the aging process because all adults age, but not all adults experience all age-associated diseases. If we are interested in aging, then life expectancy at birth may not be the most relevant information to consider because birth and childhood deaths skew the life expectancy rate dramatically downward. One way to overcome this problem is to compare the health and longevity of adult groups. For example, a comparative study carried out from 1994 to 1998 of 80 year old of members of the Church of Jesus Christ of Latter-day Saints (Mormons) and non-Mormons in Utah indicated that their remaining years of life expected were 8.2 for Mormon males, 6.5 for non-Mormon males, 10.3 for Mormon females, and 7.1 for non-Mormon females. The data suggest that the healthy lifestyle of Mormons, which includes eating nutritious food, exercising regularly, and a ban on the use of tobacco, alcohol, coffee, tea, and drugs, contributes to a healthy and long life.

The aging process is common to all living things because the phenomena of aging and death are universal. It occurs even under optimal living conditions. The aging process is under genetic control; life span and manifestations of aging differ among species and individual members of a species. The aging process is also subject to environmental influences like other biochemical reactions.

The aging process produces aging changes at an increasing rate with advancing age so that few reach 100 years and none live beyond about 122 years. The longest lived person whose date of birth can be confirmed was Jeanne Calment of France (1875–1997), who died at age 122 years, 164 days. She met Vincent van Gogh when she was 12. The longest undisputed lifespan for a male is that of Jiroemon Kimura of Japan (1897–2013), who died in June 2013 at age 116 years, 54 days.

There is considerable debate among scientists regarding how much of an increase in life expectancy is possible in the future. There is general agreement that significantly reducing old age diseases will increase life expectancy by 5–10 years, so that the average life expectancy may reach close to 100 years. Some scientists claim that 100 years will be the limit because the aging process is like a clock and when the time runs out nothing can be done. Other scientists who study the biology of aging believe that the development of interventions which slow aging is inevitable, so it should be possible to slow down the aging process, so that people can live to 150 years or more. To evaluate the possibility of a very long life span, it is necessary to consider what causes aging.

The study of aging—*gerontology*—is a relatively new science that has made incredible progress over the last 30 years. In the past, scientists looked for a single theory that explained aging. There are two main groups of aging theories. The first group states that aging is natural and programmed into the body, while the second group of aging theories says that aging is a result of damage which is accumulated over time. In both groups, the role of DNA appears to be pivotal. In the end, aging is a complex interaction of genetics, chemistry, physiology, and behavior.

Let us first consider the evolutionary significance of a limited lifespan. Evolution depends on genetic variations in populations, and for these variations to be present prior to reproduction. Genetic variations that occur after the reproductive phase of life do not contribute to evolution. Aging or otherwise purposely limited lifespan helps evolution by freeing resources for younger, and therefore, potentially better-adapted individuals to reproduce. Also, rates of evolution are increased with shorter lifespans because of the greater number of life-cycles per unit time. Microorganisms with generation times of minutes can genetically adapt thousands of times more rapidly than animals with generation times in days or years.

One of the most interesting ideas within the framework of programmed theories of aging involves *telomeres*. A telomere is a region of repetitive nucleotide sequences at each end of a chromosome that protects against combining with neighboring chromosomes. Telomeres prevent one chromosome from binding to another (DNA is sticky). The feature of telomeres that relates to aging is that during replication of chromosomal DNA, the enzymes that duplicate DNA cannot continue their duplication all the way to the end of a chromosome. This results in a shortening of chromosomes with each cell division. It is this shortening of telomeres at the ends of chromosomes that has been hypothesized to be a clock that determines cell aging.

Telomeres were discovered by Elizabeth Helen Blackburn between 1975 and 1977 while she was a postdoctoral researcher at Yale University. Blackburn has always been fascinated by how life works. Born in 1948, she grew up by the sea in a remote town in Tasmania, Australia, collecting ants from her garden and jellyfish from the beach. When she began her scientific career, she moved to biochemistry because, as she says, "It provides a deep knowledge of the smallest possible subunit of a process." Blackburn discovered telomeres by determining the DNA base sequence at the chromosome tips of a single-celled freshwater creature called *Tetrahymena.* She found a repeating DNA motif, the 6-base sequence TTGGGG repeated 20–70 times, which acts as a protective cap. The caps, dubbed telomeres, were subsequently found by Jack William Szostak of Harvard University to be present on all eukaryotic organisms, from single cell yeasts to humans. Telomeres shield the ends of our chromosomes each time our cells divide and the DNA is copied, but they wear down with each division.

In the 1980s, Blackburn and her graduate student Carol Widney Greider at the University of California, Berkeley, discovered an enzyme called telomerase that can protect and rebuild telomeres. Even so, our telomeres dwindle over time. And when they get too short, our cells start to malfunction and lose their ability to divide—a phenomenon that is now recognized as a key process in aging. For her research on how chromosomes are protected by telomeres and the enzyme telomerase, Blackburn was awarded the 2009 Nobel Prize in Physiology or Medicine, sharing it with Carol Greider and Jack Szostak (Fig. 17.1).

In 2000, Blackburn received a visit from a young psychiatrist, Elissa Epel, which changed the course of her research. Psychiatrists and biochemists don't usually have much to talk about, but Epel was interested in the damage done to the body by chronic stress, and she had a radical proposal. With some trepidation at approaching such a senior scientist, Epel asked Blackburn for help with a study of mothers going through one of the most stressful situations that she could think of—caring for a chronically ill child. Epel's plan was to ask the women how stressed they felt, then look for a relationship between their state of mind and the state of their telomeres. Epel was interested in the idea that if we look deep within cells we might be able to measure the wear and tear of stress and daily life. After reading about Blackburn's work on aging, she wondered if telomeres might fit the bill. Blackburn agreed to measure the length of

FIGURE 17.1 The 2009 Nobel laureates in Physiology or Medicine. From left to right: Elizabeth Blackburn, Jack Szostak, and Carol Greider. *(From ®© The Nobel Foundation.)*

telomeres and the levels of the telomerase enzyme in individuals at different degrees of stress.

The results of the Epel/Blackburn experiment, published in 2002, provided the first evidence that psychological stress—both perceived stress and chronicity of stress—is significantly associated with lower telomerase activity and shorter telomere length. Women with the highest levels of perceived stress have telomeres shorter on average by the equivalent of at least one decade of additional aging compared to low stress women. Recently, it has been shown that patients with bipolar disorder, psychosis, major depression, and anxiety disorders have shorter telomeres than healthy people of the same age. It has also been reported that shorter telomeres reflect premature aging in psychiatric patients. Interestingly, long-term lithium therapy leads to a detectable relative increase in telomere length versus controls. The effect of stress on telomere length and aging is not limited to humans. The stress of birds' continent-spanning annual migrations leads to faster aging and earlier death. DNA analysis revealed that telomere structures on the ends of migratory birds are shorter than on their nonmigratory counterparts.

Conventional medical tests give us our risk of particular conditions—high cholesterol warns of impending heart disease, for example, while high blood sugar predicts diabetes. Telomere length, by contrast, gives an overall reading of how healthy we are: our biological age. And although we already know that we should exercise, eat well, and reduce stress, many of us fall short of these goals. Blackburn believes that putting a concrete number on how we are doing could provide a powerful incentive to change our behavior. In fact, she and Epel have just completed a study showing that simply being told their telomere length caused volunteers to live more healthily over the next year than a similar group who weren't told.

There is evidence both for and against the telomere hypothesis of aging. The arguments in favor of the hypothesis include the following: telomeres are shorter in most tissues from older individuals compared to younger individuals; telomeres in normal cells from young individuals progressively shorten when grown in the laboratory; experimental elongation of telomeres by introducing telomerase extends the age that cells can multiply in the laboratory; children born with progeria (early aging syndrome) have shortened telomeres compared with age-matched normal children.

The arguments against the hypothesis are as follows: no connection exists between mean telomere length of cells and longevity of different mammalian species. Of all studied primates, humans appear to have the shortest telomeres and the longest lifespan; the association of telomere lengths with age is a very weak correlation applicable to populations but not necessarily to individuals (a significant number of older people have much longer telomeres than those in a significant portion of the younger people); as people age, telomere lengths can get longer as well as shorter (inevitable telomere shortening due to cell division is simply not the case); telomere length is not a particularly good biomarker for predicting the life expectancy of old people.

In conclusion, telomere shortening probably contributes to cell aging, but the process of aging is much more complicated than that and many other factors are involved including environmental cell damage.

Error theories of aging assert that chemical reactions going on continuously throughout the cells and tissues constitute the aging process or are major contributors to it. In mammalian systems the major damaging reactions involve oxygen. Oxygen plays an ambivalent role in nature. Human beings, animals, and plants require oxygen for the production of energy from nutrients by respiration. We can only survive a few minutes without oxygen. However, the oxygen we inhale is not completely harmless to us. The danger lies in the fact that oxygen molecules O_2 form a small amount of free radicals during respiration, such as hydroxyl radicals ($^{\cdot}OH$) and superoxide radicals $\cdot O_2^-$. The dot next to the oxygen represents a single electron. Compounds containing single unpaired electron are referred to as radicals, and they are very reactive. They attack all sorts of important biological molecules in our tissues (DNA, RNA, and proteins, among others), which results in the process of aging, and eventually this accumulated damage results in death of the organism.

The free radical theory of aging provides reasonable explanations for a number of age-associated phenomena, including the relationship of the average life spans of animals to their metabolic rates. For example, large mammals, such as elephants and humans, live longer and have a slower rate of metabolism (generate less oxygen radicals) than small mammals, such as mice and guinea pigs. Food restriction, which is expected to lower the rate of metabolism and production of free radicals, has been shown to increase the life span of mice, rats, fruit flies, nematodes, and rotifers. Rats and mice on a calorie restricted diet (30% fewer daily calories) live up to 40% longer. The generally longer lifespans of females compared to males correlates with a lower level of oxygen radicals in females than males. It has been suggested that the female sex hormone estrogen, which is also an antioxidant, that is, it neutralizes naturally occurring oxygen radicals, contributes to reducing a woman's risk of aging and dying. Estrogen therapy has also been shown in some studies to delay the onset of Alzheimer's disease (Fig. 17.2).

The DNA damage theory of aging proposes that aging is a consequence of unrepaired accumulation of naturally occurring DNA damage. In the previous chapter, I discussed how DNA mutations in proto-oncogenes can lead to uncontrolled growth and cancer. DNA damage is different. It does not involve changes in the base sequence of DNA, but rather a DNA alteration that leads to an abnormal structure of DNA. For example, oxygen radicals can cause a change in the chemical structure of the bases in DNA and can also cause single strand breaks in the deoxyribose-phosphate backbone of DNA. Nonreplicating cells, including brain neurons and muscle cells, accumulate DNA damage with time that likely contributes to aging. Single strand breaks and other damage to DNA block the transcription of messenger RNAs from DNA. This would interfere with the synthesis of the proteins coded for by the genes in which the blockages occurred.

In humans and other mammals, DNA damage occurs frequently and DNA repair processes have evolved to compensate. In estimates made for mice, on average approximately 4000 DNA lesions occur per hour in each mouse cell. In any cell some DNA damage may remain despite the action of repair processes. DNA damage in replicating cells is a major cause of mutations and cancer, whereas the accumulation of unrepaired DNA damage in nonreplicating or

FIGURE 17.2 **Aging is affected by the food you eat.**

infrequently dividing cells leads to a decline in cell function and is a prominent cause of aging.

The mitochondrial theory of aging, a variant of free radical theory of aging, proposes that accumulation of damage to mitochondria leads to aging of humans and animals. As discussed in Chapter 14, mitochondria are fundamental structures found in the cytoplasm of all eukaryotic cells. Mitochondria use oxygen to generate chemical energy, in the form of adenosine triphosphate (ATP). Mitochondria resemble bacteria in many ways. Mitochondria have their own bacterial-like DNA, which they duplicate when they divide, similarly to bacteria, into new mitochondria. It has been shown that mitochondrial function declines and mitochondrial DNA mutations increase in tissue cells in an age-dependent manner. As they age, mitochondria become less efficient at turning fuel into energy. On the other hand, aging mitochondria become more "productive" in a negative way: they start producing more oxygen free radicals. The effect of aging mitochondria is a body operating at one-half to one-fourth the energy it had at youth. The brain is perhaps the most important organ affected by aging, since it consumes more energy than any other organ of the body. An energy deficit in the brain and central nervous system affects the activities of all organs throughout the body as well as mental acuity and mood.

Another error process that contributes to aging is glycation, a nonenzymatic reaction between sugars, such as glucose and fructose, and proteins, lipids, or nucleic acids. The result is that sugars becomes connected to important biopolymers and prevent their proper functioning. It is the same reaction, which in cooking gives rise to the browning of various meats like steak, when seared and grilled, or the golden-brown color of French fries. Attachment of sugars to proteins causes the proteins to bind to each other and accumulate in a specific tissue, such as cartilage, lungs, arteries, and tendons. Excess sugars in the blood stream can cause protein molecules to literally stick together. Basically, things become stiffer. When tissues stiffen, they do not function as efficiently. Many of the symptoms of aging have to do with the stiffening of tissues. Cataracts, for example, are a stiffening of the lenses of your eye. People with diabetes, who have elevated blood sugar, develop senescence-associated disorders much earlier than the general population, but can delay such disorders by rigorous control of their blood sugar levels.

Progeria is an extremely rare genetic disorder wherein symptoms resembling aspects of aging are manifested at a very early age. The word progeria comes from the Greek words "pro" (*πρό*), meaning "before" or "premature," and "gēras" *γῆρας*, meaning "old age." The disorder has a very low incidence rate, occurring in an estimated one per eight million live births. Those born with progeria typically live to their midteens to early 20s. It is a genetic condition that occurs as a new mutation, and is rarely inherited, as those with the disease usually do not live to reproduce. The mutation, which appears to occur in the sperm prior to conception, is a single base replacement of cytosine by thymine in a gene that provides instructions for making a protein called lamin A. This protein plays an important role in determining the shape of the nucleus within cells. Mutations that cause progeria result in the production of an abnormal version of the lamin A protein. The altered protein makes the nuclear envelope unstable and progressively damages the nucleus, making cells more likely to die prematurely. Sufferers exhibit symptoms resembling accelerated aging, including wrinkled skin. Scientists are particularly interested in progeria because it might reveal clues about the normal process of aging.

Life According to Sam was a 2013 documentary on a progeria patient Sam Berns, who died of the disease on January 10, 2014, age 17. He was a fan of the New England Patriots football team, and had he lived another day, he would have served as the team's honorary captain in their playoff game versus the Indianapolis Colts. The film explains progeria and follows the process of trying to find a cure for it. In an interview, Sam Berns had said that the most important thing people should know about him is that he had a very happy life.

Considering what we now know about the process of aging, what can we do to slow down the process and live a longer and healthier life? Probably no one answered this question more clearly than Bruce Ames of the University of California at Berkeley, when he wrote:

> *Eat a good diet. And then don't smoke, because smoking takes about 8 or 10 years off your life. And I think bad diets are probably another 8 or 10 years off your life, though we don't know the exact number. But I suspect it's going to be even worse than smoking. And the costs are huge. You have to convince people that they're going to lead miserable lives if they get fat, have years suffering from diabetes and their brain will be all fogged. The choice seems obvious to me.*

As I have discussed in this chapter, oxygen free radicals, formed when the body changes oxygen and food into energy, is a major source of DNA damage and aging. In principle, antioxidants should slow down aging. However, so far, studies of antioxidant supplements in humans have yielded little support for this conclusion. Further research, including large-scale epidemiological studies, might clarify whether dietary antioxidants can help people live longer, healthier lives. For now, the American Diabetes Association does not advise routine supplementation with antioxidants, such as vitamins E and C and carotene because of lack of evidence of efficacy and concern related to long-term safety. Although the effectiveness of dietary antioxidant supplementation remains controversial, there is positive evidence for the health benefits of fruits and vegetables. It may be that combinations of nutrients found in foods have greater protective effects than each nutrient taken alone.

There is strong evidence from animal and human studies that natural antioxidants in colorful vegetables and fruits, such as leafy greens, red tomatoes, blueberries, and carrots, help stop free radical molecules from damaging healthy cells. The National Institute on Aging (United States of America) recommends that adults eat five to nine servings of fruit and vegetables a day. Omega-3 fatty acids in fatty fish, such as salmon, lake trout or tuna, offer many antiaging benefits. Native Inuit of Alaska eat a diet very high in whale, seal, and salmon, and are remarkably free of heart disease, rheumatoid arthritis, and diabetes. Clearly, maintaining a low level of glucose in the blood will reduce DNA and protein damage due to glycation.

For years, scientists have been puzzled by the so-called "French Paradox," which refers to the fact that French people have a relatively low incidence of coronary heart disease, while having a diet relatively rich in saturated fats. Observing this paradox, many scientists have suggested it is the regular and moderate consumption of red wine that might be protecting against cardiovascular disease. (The operative word here is moderate; drinking an excessive amount of alcohol has well-documented negative effects on health.) Some studies in animals have suggested that a powerful antioxidant in grapes and red wine, called resveratrol, will lower the risk of cancer, heart disease, and premature aging. Resveratrol administration has been shown to increase the lifespans of yeast, worms, fruit flies, fish, and mice fed a high-calorie diet, but it is not known whether resveratrol will have similar effects in humans. New research outlining the biological pathways that activate an antiinflammatory effect in human cells could provide a mechanism for the suggested benefits of resveratrol. It should be noted that smelling is an important part of consuming wine. It is through the aromas of wine that wine is tasted. Furthermore, resveratrol and other volatile components in wine are more efficiently absorbed through the nose than the digestive tract where they are rapidly broken down.

Some biologists have suggested that increasing the length of telomeres at the end of chromosomes would inhibit aging and lengthen life. They argue that since increasing the activity of telomerase and thereby lengthening telomeres makes cancer cells immortal, it should prevent normal cells from aging? Could we extend lifespan by preserving or restoring the length of telomeres with telomerase? If so, would that increase our risk of getting cancer? Scientists are not yet sure. But they have been able to use telomerase in the laboratory to keep human cells dividing far beyond their normal limit, and the cells do not become cancerous.

One study showed that when people are divided into two groups based on telomere length, the half with longer telomeres lives an average of 5 years longer than those with shorter telomeres. This study suggests that lifespan could be increased 5 years by increasing the length of telomeres in people with shorter ones. People with longer telomeres still experience telomere shortening as they age. One could speculate that by completely stopping telomere shortening, it might be possible to add 10 years and perhaps 30 years to our life span.

In conclusion, reducing oxidative damage, glycation, and telomere shortening have the potential to reduce the rate of aging. How much this will increase life expectancy remains to be seen. What Winston Churchill said in 1942 in reference to the Second World War also applies to our current understanding of aging:

> *Now this is not the end. It is not even the beginning of the end. But it is, perhaps, the end of the beginning.*

NOTES AND REFERENCES

168 Demeny, G., McNicoll, G., 2003. Encyclopedia of Population. McMillian Publishing Group, London.

169 Merrill, R.M., 2004. Life expectancy among LDS and non-LDS in Utah. Demogr. Res. 10, 61–82.

169 Young, M.C. (Ed.), 1997. The Guinness Book of Records. Bantam Books, New York, NY.

170 Blackburn, E., Szostak, J., 1984. The molecular structure of centomeres and telomeres. Ann. Rev. Biochem. 53, 163–194.

170 Lindsey, J., et al., 1991. In vivo loss of telomeric repeats with age in humans. Mutat. Res. 256, 45–48.

170 Bodnar, A.G., et al., 1998. Extension of lifespan by introduction of telomerase in normal human cells. Science 279, 349–352.

171 Steinert, S., et al., 2002. Telomere biology and cellular aging in nonhuman primate cells. Exp. Cell Res. 272, 146–152.

171 Epel, E.S., et al., 2004. Accelerated telomere shortening in response to life stress. Proc. Natl. Acad. Sci. USA 101, 17312–17315.

171 Corbett, N., Alda, M., 2015. On telomeres long and short. J. Psychiatry Neurosci. 40, 3–4.

173 Recently, the free radical theory has been referred to as the oxidative-stress theory of aging because not all the damaging reactive oxygen molecules are free radicals.

174 Jacobs, E.G., et al., 2013. Accelerated cell aging in female APOE-ε4 carriers: implications for hormone therapy use. PLoS One 8 (2), e54713.

174 Gkogkolou, P., Böhm, M., 2012. Advanced glycation end products: key players in skin aging? Dermato-endocrinology 4, 259–270.

176 He, F.J., et al., 2007. Increased consumption of fruit and vegetables is related to a reduced risk of coronary heart disease: meta-analysis of cohort studies. J. Hum. Hypertens. 21, 717–728.

Appendix

Nobel Prizes Awarded for Nucleic Acid Research

Emil Fischer (1902) "in recognition of the extraordinary services he has rendered by his work on sugar and purine syntheses"

Albrecht Kössel (1910) "in recognition of the contributions to our knowledge of cell chemistry made through his work on proteins, including the nucleic substances"

Thomas Morgan (1933) "for his discoveries concerning the role played by the chromosome in heredity"

Hermann Muller (1946) "for the discovery of the production of mutations by means of X-ray irradiation"

Lord Alexander Todd (1957) "for his work on nucleotides and nucleotide coenzymes"

George Beadle and Edward Tatum (1958) "for their discovery that genes act by regulating definite chemical events"

Joshua Lederberg (1958) "for his discoveries concerning genetic recombination and the organization of the genetic material of bacteria"

Severo Ochoa and Arthur Kornberg (1959) "for their discovery of the mechanisms in the biological synthesis of ribonucleic acid and deoxyribonucleic acid"

Francis Crick, James Watson, and Maurice Wilkins (1962) "for their discoveries concerning the molecular structure of nucleic acids and its significance for information transfer in living material"

François Jacob, André Lwoff, and Jacques Monod (1965) "for their discoveries concerning genetic control of enzyme and virus synthesis"

Robert Holley, Har Gobind Khorana, and Marshall Nirenberg (1968) "for their interpretation of the genetic code and its function in protein synthesis"

Max Delbrück, Alfred Hershey, and Salvador Luria (1969) "for their discoveries concerning the replication mechanism and the genetic structure of viruses"

It's in Your DNA. http://dx.doi.org/10.1016/B978-0-12-812502-1.00019-6

David Baltimore, Renato Dulbecco, and Martin Temin (1975) "for their discoveries concerning the interaction between tumour viruses and the genetic material of the cell"

Werner Arber, Daniel Nathans, and Hamilton Smith (1978) "for the discovery of restriction enzymes and their application to problems of molecular genetics"

Paul Berg (1980) "for his fundamental studies of the biochemistry of nucleic acids, with particular regard to recombinant-DNA"

Walter Gilbert and Frederick Sanger (1980) "for their contributions concerning the determination of base sequences in nucleic acids"

Aaron Klug (1982) "for his development of crystallographic electron microscopy and his structural elucidation of biologically important nucleic acid–protein complexes"

Barbara McClintock (1983) "for her discovery of mobile genetic elements"

Sidney Altman and Thomas Cech (1989) "for their discovery of catalytic properties of RNA"

J. Michael Bishop and Harold E. Varmus (1989) "for their discovery of the cellular origin of retroviral oncogenes"

Kary Mullis (1993) "for his invention of the polymerase chain reaction (PCR) method"

Richard Roberts and Phillip Sharp (1993) "for their discoveries of split genes"

Roger Kornberg (2006) "for his studies of the molecular basis of eukaryotic transcription"

Andrew Fire and Craig Mello (2006) "for their discovery of RNA interference—gene silencing by double-stranded RNA"

Venkatraman Ramakrishnan, Thomas Steitz, and Ada Yonath (2009) "for studies of the structure and function of the ribosome"

Elizabeth H. Blackburn, Carol W. Greider, and Jack W. Szostak (2009) "for the discovery of how chromosomes are protected by telomeres and the enzyme telomerase"

John B. Gurdon and Shinya Yamanaka (2012) "for the discovery that mature cells can be reprogrammed to become pluripotent"

Glossary of Scientific Terms

To increase understanding is a laudable goal, hence the definitions and explanations below.

—Frank Herbert, author of *Dune*

Exactness cannot be established in the arguments unless it is first introduced into the definitions.

—Jules Henri Poincaré, 19th century mathematician

Adenine a purine base that is present in DNA and RNA.

Adenosine triphosphate (ATP) a molecule consisting of adenine, ribose, and three phosphate groups. It is the key energy-rich compound in cells, that is, when it is broken down a large amount of energy is released.

Aerobic cells those that utilize oxygen.

Agar a polysaccharide derived from seaweed; it is used widely in microbiology laboratories as a solidifying agent in growth media.

Alkaptonuria a hereditary disease in which the patient lacks the enzyme for degrading homogentisic acid and thus excretes it in the urine.

Ames test a widely employed method that uses bacteria to test whether a given chemical can cause cancer. The test measures the ability of the chemical to cause mutations in bacteria.

Amino acid a group of important organic compounds that which comprises the building blocks from which proteins are constructed. There are 20 common and two rare amino acids that are found in proteins.

Amniotic membrane the innermost layer of the placenta consisting of a thick basement membrane.

Antibody a blood protein produced in response to and counteracting a specific antigen.

Antigen foreign substance that induces an immune response in the body, especially the production of antibodies.

Archaea one of the three domains of life (the other are Bacteria and Eukaryota). Archaea are prokaryotes that often inhabit extreme environments.

Bacteriophages (phages) viruses that infect and multiply in bacteria.

Cancer a group of diseases characterized by uncontrolled cellular duplication.

Carcinogen a substance or agent that causes cancer.

Catalyst an agent that increases the rate of a reaction, itself emerging unchanged at the end of the process. Since it is continually regenerated, a small amount of catalyst can produce a large increase in rate. In living organisms the important catalysts are large protein molecules called enzymes.

Chemical evolution the theory, initially proposed by Oparin, which states that life arose gradually from nonliving material. The theory emphasizes the need for a long series of chemical changes as a prerequisite to the formation of life.

It's in Your DNA. http://dx.doi.org/10.1016/B978-0-12-812502-1.00020-2

Chloroplast membrane bound structures found in green plant cells; they are the sites of photosynthesis.

Chromosomes threadlike structures containing DNA and protein which are present in the nucleus of all animals and plant cells. They constitute the heredity organelles of the cell.

Codon a sequence of three adjacent nuceotides that code for an amino acid or chain termination of proteins.

Commensalism symbiosis where the association is advantageous to one and doesn't affect the other(s).

Congenital relating to a condition present at birth, whether inherited or caused by the environment, especially the uterine environment.

Coprophagia the consumption of feces. The word is derived from the Greek *κόπροζ* copros, "feces" and *φαγεῖν* phagein, "to eat."

CRISPR (*C*lustered *R*egularly *I*nterspaced *S*hort *P*alindromic *R*epeat) segments of prokaryotic DNA containing short repetitions of base sequences. Each repetition is followed by short segments of "spacer DNA" from previous exposures to a bacteriophage virus or plasmid.

Cyanobacteria photosynthetic bacteria (prokaryote), generally blue-green in color and in some species capable of nitrogen fixation.

Cytogenetics the study of inheritance in relation to the structure and function of chromosomes.

Cytology the branch of life science which deals with the study of cell structures.

Cytosine a pyrimidine base that is present in DNA and RNA.

Density in science, the degree of compactness of a substance, its mass per unit volume.

DNA deoxyribonucleic acid, a self-replicating polymer that carries genetic information in animals, plants, microorganisms, and most viruses. It is present in cells as two paired complementary strands in the form of a double helix. Each strand is a long chemical chain made up of four nucleotides. Each nucleotide consists of three components—the sugar deoxyribose, phosphate, and one of the four nitrogen bases, adenine (A), guanine (G), thymine (T), and cytosine (C). The sequence of the nitrogen bases determines the sequence of RNAs (transcription) and proteins (translation).

DNA fingerprinting a technique used especially for identification (as for forensic purposes) by extracting and identifying the base-pair pattern in an individual's DNA—called also DNA typing.

DNA polymerase an enzyme that catalyzes the formation of DNA from deoxyribonucleoside triphosphates utilizing existing DNA as a template.

DNase deoxyribonuclease, an enzyme that catalyzes the degradation of DNA.

Ecology the branch of science concerned with the interrelationship of organisms and their environments.

Endophyte a microorganism that that lives within a plant cell.

Endosymbiont hypothesis postulates that early eukaryotic cells lacking mitochondria and chloroplasts incorporated aerobic prokaryotes and, rather than digesting them, formed a symbiotic relationship with them, offering them nutrients and shelter (therefore the "endo-" part) and getting efficient energy generating systems in return.

Endosymbionts referring to microorganisms living inside host cells.

Enzymes proteins which act as catalysts. Catalysis by enzymes is characterized by great efficiency and high specificity. Most enzymes catalyze only one type of biochemical reaction.

Epidemiology the branch of medicine dealing with the incidence and prevalence of disease in populations and with the detection of possible sources and causes of disease epidemics.

Escherichia coli (E. coli) a bacterium found in the intestines of humans; it is easy to grow and manipulate in the laboratory and thus has been used widely in the study of molecular biology.

Estrogen female sex hormones produced primarily by the ovarian follicles of female mammals, capable of inducing estrus, developing and maintaining secondary female sex characteristics, and preparing the uterus for the reception of a fertilized egg.

Eukaryotic cell typical of all cell types except bacteria, consisting of a well-defined nucleus separated from the cytoplasm by a nuclear membrane and a structurally differentiated cytoplasm.

Evolution the gradual development of something, especially from a simple to a more complex form. In biology, the process by which different kinds of living organisms are thought to have developed and diversified from earlier forms during the history of the earth.

Exosymbionts referring to microorganisms living outside host cells.

Fetus a human being or animal in the later stages of development before it is born.

Fitness the propensity of an organism to survive and reproduce in a specified environment and population.

Forensics relating to the use of scientific knowledge or methods in solving crimes.

Free radical An atom or group of atoms that has at least one unpaired electron and is therefore unstable and highly reactive. In animal tissues, free radicals can damage cells and are believed to accelerate the progression of cancer, cardiovascular disease, and age-related diseases.

French paradox refers to the fact that French people have a relatively low incidence of coronary heart disease, while having a diet relatively rich in saturated fats.

Gene a unit of hereditary material located on the chromosome that is transferred from a parent to offspring and is held to determine some characteristic of the offspring. In modern terms a gene can be considered a segment of DNA carrying the information for a single RNA and protein molecule. The word gene was coined by the Danish botanist Wilhelm Johannsen.

Genetic code the four nitrogenous bases (adenine, thymine, guanine, and cytosine) in DNA constitute an alphabet of four letters; the sequence of these bases in DNA and subsequently in messenger RNA forms a code which contains the information for synthesizing specific proteins. The code consists of three-letter words, each of which is the code for a specific amino acid. In this way the sequence of amino acids in proteins is determined by the sequence of bases in DNA.

Genetic engineering the group of applied techniques of genetics and biotechnology used to cut up and join together genetic material and especially DNA from one or more species of organism and to introduce the result into another organism in order to change one or more of its characteristics.

Genome The genome is the complete set of genetic information for an organism, encoded as DNA sequences within the chromosome pairs in cell nuclei and in a small DNA molecule found within individual mitochondria.

Genotype the internally coded, inheritable information carried by all living organisms. This stored information is used as a "blueprint" or set of instructions for building and maintaining a living creature.

Germ cell a reproductive cell, either the egg or the sperm cell. Each mature germ cell is haploid, meaning that for humans it has a single set of the 23 chromosomes and contains half the usual amount of DNA and half the usual number of genes.

Germ-free animals a laboratory animal born and raised under sterile conditions, free of exposure to microorganisms. The diet is controlled, preventing exposure to microorganisms that may be in food.

Gerontology the scientific study of old age, the process of aging, and the particular problems of old people.

Glycation (sometimes called nonenzymatic glycosylation) is a chemically reaction in which a sugar molecule, such as glucose or fructose, become connected to a protein or lipid molecule, without the controlling action of an enzyme.

Group selection refers to a mechanism of evolution in which natural selection acts at the level of the group instead of at the level of the individual.

Guanine a purine base that is present in DNA and RNA.

Holobiont = metaorganism a host organism (plant or animal) with all associated microorganisms, including bacteria, protists, and viruses.

Hologenome the sum of the genetic information of the host and all its microbiota, including viruses.

Hologenome concept of evolution the proposal that the holobiont with its hologenome is a level of selection in evolution.

Homogentisic acid an organic compound which is formed in the body from the amino acids phenylalanine and tyrosine. Patients with the disease alkaptonuria cannot degrade this compound and thus excrete it in their urine.

Horizontal gene transfer (HGT) transfer of genetic material to different places on the chromosome and between different organisms in a manner other than traditional vertical transfer of genes from the parental generation to offspring via sexual or asexual reproduction.

Human genome the complete set of genetic information for humans. This information is located as DNA sequences within the 23 chromosome pairs in cell nuclei and in a small DNA molecule found within individual mitochondria. The total number of human genes is approximately 22,000.

Hygiene hypothesis a theory that suggests a young child's environment can be "too clean" to effectively stimulate or challenge the child's immune system to respond to various threats during the time a child's immune system is maturing.

Immunotherapy the prevention or treatment of disease with substances that stimulate the immune response.

Insulin a protein hormone produced in the pancreas and used in the metabolism of sugar and other carbohydrates.

In vitro (Latin: in glass): pertaining to experiments performed with cell-free extracts.

In vivo (Latin: in life): pertaining to experiments performed with intact living cells.

Lamarckism the idea that an organism can pass on characteristics that it acquired during its lifetime to its offspring (also known as heritability of acquired characteristics).

Leukemia cancer of the body's blood-forming tissues, including the bone marrow and the lymphatic system.

Lichen a symbiotic organism that is produced from an algae or cyanobacteria (or both) living together with a fungus.

Macromolecule a large molecule, usually built from small units, for example, proteins, nucleic acids, and polysaccharides.

Mating preference selection or choice of sexual partner in animals; often this reproductive preference is based on traits in the potential mate, such as coloration, size, and smell.

Meiosis a type of cell division that results in daughter cells each with half the number of chromosomes of the parent cell, as in the production of gametes and plant spores.

Melanoma a highly malignant tumor that starts in melanocytes of normal skin or moles and metastasizes rapidly and widely.

Messenger RNA (mRNA) RNA manufactured in the nucleus, which moves into the cytoplasm, attaches to ribosomes, and serves as a template for protein synthesis.

Metaorganism see holobiont.

Microbiome = microbiota the sum of the associated microorganisms in a particular environment, especially the body or a part of the body of an animal or plant. For example, human gut microbiota and the human gut microbiome refer to all the microorganisms present in the human gut.

Mitochondrion a structure found in the cytoplasm of aerobic animal and plant cells. It is the major site of ATP production.

Multilevel selection the theory that natural selection can operate simultaneously at different levels of the biological hierarchy, such as the gene, individual, group, and holobiont.

Mutagen an agent, such as radiation or a chemical substance, which increases the frequency of mutation.

Mutation any sudden, heritable change in the structure of DNA caused by the alteration of single base units, or the deletion, insertion, or rearrangement of larger sections of genes or chromosomes. Mutations can be silent, not affecting any function of the cell, or can result in a change in function or structure.

Mutualism symbiosis where both the host and the symbiont benefit from the interaction.

Mycorrhiza fungi that grow in association with the roots of a plant in a symbiotic relationship.

Neurospora crassa the common bread mold; it is especially useful for biochemical and genetic studies.

Next-generation sequencing also known as high-throughput sequencing, is the catch-all term used to describe a number of different modern sequencing technologies. These recent technologies allow us to sequence DNA much more quickly and cheaply than the previously used Sanger sequencing method.

Nitrogen fixation The conversion of atmospheric nitrogen gas into ammonia by biological or industrial processes.

Nitrogenous bases Molecules composed of rings of carbon and nitrogen atoms. Important nitrogenous bases in the cell are the purines and pyrimidines in DNA and RNA.

Nucleic acid a polymer of nucleotides (see DNA and RNA).

Nuclein the initial term to describe nucleic acid.

Nucleotide one of the nitrogen bases joined to a sugar which is also connected to a phosphate group (base + sugar + phosphate). It is the monomeric unit of nucleic acids.

Nucleus that part of the cell which contains the genetic material. The nucleus is bounded by a nuclear membrane.

Oncogene a cancer-causing or cancer-promoting gene.

Organic molecules compounds containing carbon that are typically found in living systems.

Panspermia the theory that life on Earth arose from "seeds" (spores) which constantly bombard our planet.

Parasitism (or pathogenesis) symbiosis where the symbiont benefits and the host suffers damage.

PCR polymerase chain reaction, an in vitro technique for rapidly synthesizing large quantities of a given DNA segment that involves separating the DNA into its two complementary strands, using DNA polymerase to synthesize two-stranded DNA from each single strand, and repeating the process many times.

Phages see bacteriophages.

Phenotype any observable characteristic of an organism, such as its morphology, development, biochemical, or physiological properties, or behavior. Phenotypes are influenced by a combination of genetic and environmental factors.

Photosynthesis the enzyme-catalyzed conversion of light energy into useful chemical energy and use of the chemical energy to form sugars and oxygen from carbon dioxide and water.

Phyllosphere the upper parts of the plant (leaves, stems, flowers, and fruit).

Placenta a flattened circular organ in the uterus of pregnant mammals, nourishing and maintaining the fetus through the umbilical cord.

Plasmid a DNA molecule that is separate from and can replicate independently of the chromosomal DNA. They are double stranded and, in many cases, circular. Plasmids usually occur naturally in bacteria, but are sometimes found in eukaryotic organisms.

Polymer a large molecule made up of regular subunits termed monomers. For example, DNA (polymer) made up of nucleotides (monomers).

Polyploidy an organism or cell having more than twice the haploid number of chromosomes.

Polysaccharide large carbohydrate molecules composed of many sugars joined together; for example, starch and glycogen.

Probiotics live microorganisms which, when administered in adequate amounts, confer a health benefit on the host.

Progeria an extremely rare, progressive genetic disorder that causes children to age rapidly.

Prokaryote a microscopic single-celled organism that has neither a distinct nucleus with a membrane nor other specialized organelles. Prokaryotes include the bacteria and cyanobacteria.

Protein polymers of amino acids that perform the bulk of cellular functions, such as the catalysis of biochemical reactions (enzymes) and formation of structural elements of cells and tissues that give shape to cells and power cell movement.

Proteinase an enzyme that breaks down proteins.

Proto-oncogene a normal cell gene that can become an oncogene by mutation.

Pure culture a population of cells that contains a single kind of microorganism and that has originated from a single cell.

Purine molecule consisting of two fused rings of five carbon and four nitrogen atoms. Two of the purines, adenine and guanine, serve as nitrogen bases in DNA and RNA.

Pyrimidine molecule consisting of a single ring of four carbon and two nitrogen atoms. Two of the pyrimidines, cytosine and thymine, serve as nitrogen bases in DNA and two, cytosine and uracil, serve in RNA.

Radical see free radical.

Recombinant DNA technology the group of techniques used to cut up and join together genetic material, especially DNA from different biological species, and to introduce the resulting hybrid DNA into an organism in order to form new combinations of heritable genetic material and thereby change one or more of its characteristics.

Respiration a process in living organisms involving the production of energy, typically with the intake of oxygen and the release of carbon dioxide from the oxidation of organic substances.

Resveratrol an antioxidant found in red wine.

R-factor a plasmid that codes for antibiotic resistance. Often, R-factors code for more than one antibiotic resistance factor: genes that encode resistance to unrelated antibiotics may be carried on a single R-factor.

Rhizosphere the region of the soil in contact with the roots of a plant.

Ribosome cytoplasmic particles consisting of RNA and protein; it is the site of protein synthesis.

RNA ribonucleic acid, a polymer that is chemically similar to DNA but contains the sugar ribose instead of deoxyribose and the nitrogen base uracil instead of thymine. It is usually in the form of a single strand. Its main function in cells is to facilitate protein synthesis. Three types of RNA are involved in protein synthesis: messenger RNA (mRNA) acts as an "intermediate" messenger, carrying DNA information from the nucleus to ribosomal RNA (rRNA)—containing ribosomes in the cytoplasm where the protein is synthesized. Smaller transfer RNA (tRNA) carries amino acids to ribosomes where they are polymerized to proteins.

RNA World Hypothesis a conceptual era in Earth's early evolution in which life began as biologically active, self-replicating molecules—catalytic RNA.

Roundup the trade name for a broad-spectrum systemic herbicide used to kill weeds. The active ingredient of the Roundup is glyphosate (*N*-(phosphonomethyl) glycine).

Rumen the first and largest division of the stomach in ruminant animals, in which the food is fermented by microorganisms.

Sanger sequencing a method for determining the sequence of bases in DNA developed by Frederick Sanger based on the selective incorporation of chain-terminating dideoxynucleotides by DNA polymerase during in vitro DNA replication.

Scrotum a part of the external male genitalia located behind and underneath the penis. It is the small muscular sac that contains and protects the testicles.

Single Nucleotide Polymorphism (SNP) SNP (pronounced snip) is a DNA sequence variation occurring commonly within a population (e.g., 1%) in which a single nucleotide—A, T, C, or G—in the genome differs between members of a biological species.

Somatic cell any cell of a living organism other than the reproductive cells.

Speciation the formation of new and distinct species in the course of evolution.

Species a group of living organisms consisting of similar individuals capable of exchanging genes or interbreeding.

Splicing (RNA) in molecular biology and genetics, splicing is a modification of the nascent premessenger RNA transcript in which sequences are removed and other sequences are joined. For nuclear encoded genes, splicing takes place within the nucleus after or concurrently with transcription.

Spore a thick-walled cell capable of surviving adverse environmental conditions.

Symbiosis the living together of different biological species.

Syncytin a captive retroviral envelope protein involved in human placental morphogenesis.

Taxonomy branch of science concerned with classification of organisms.

Telomere a region of repetitive nucleotide sequences at each end of a chromosome that protects against fusion with neighboring chromosomes.

Thermophilic bacteria bacteria that thrive at relatively high temperatures, between 50 and 110°C. They are suggested to have been among the earliest bacteria.

Thymine a pyrimidine base found in DNA.

Transcription a step in protein synthesis in which part of the DNA is used as a template for the production of a complimentary sequence of bases in an RNA chain.

Transformation the genetic change in a bacterium brought about by absorbing DNA from other strains of bacteria.

Translation the steps in protein synthesis which take place on the ribosome whereby information in messenger RNA is utilized to direct the synthesis of a specific protein.

Transposase an enzyme coded for by a gene in a transposon that binds to the ends of a transposon and catalyzes the movement of the transposon to another part of the genome by a cut and paste mechanism.

Transposon small, mobile DNA sequences that can replicate and insert copies at random sites within chromosomes. They have nearly identical sequences at each end, oppositely oriented (inverted) repeats and code for the enzyme, transposase, which catalyses their insertion.

Uracil a pyrimidine base found in RNA.

Variation genetic variation refers to diversity in gene frequencies. Genetic variation can refer to differences between individuals or to differences between populations.

Variegated having patches, stripes, or marks of different colors.

Virus a small microorganism that is incapable of reproducing by itself, but capable of multiplying once it enters a cell. Viruses come in diverse forms including both DNA and DNA viruses, coated with proteins and sometimes by an outer membrane made of lipids and proteins.

Selected Bibliography

Aldridge, S., 1998. The Thread of Life: The Story of Genes and Genetic Engineering. Cambridge University Press, Cambridge.

Allen, G.E., 2000. Morgan, Thomas Hunt. American National Biography. Oxford University Press, Oxford.

Andersson, M., 1994. Sexual Selection. Princeton University Press, Princeton, NJ.

Arking, R., 2006. The Biology of Aging: Observations and Principles. Oxford University Press, Oxford.

Arora, N.K. (Ed.), 2013. Plant Microbe Symbiosis: Fundamentals and Advances. Springer, Heidelberg.

Avise, J.C., 2004. The Hope, Hype & Reality of Genetic Engineering: Remarkable Stories From Agriculture, Industry, Medicine, and the Environment. Oxford University Press, New York.

Bannister, R.C., 1989. Social Darwinism: Science and Myth in Anglo-American Social Thought. Temple University Press, Philadelphia.

Barrangou, R., van der Oost, J. (Eds.), 2013. CRISPR-Cas Systems: RNA-mediated Adaptive Immunity in Bacteria and Archaea. Springer, Berlin, Heidelberg.

Bearn, A.G., 1993. Archibald Garrod and the Individuality of Man. Oxford Press, Oxford.

Blaser, M.J., 2014. Missing Microbes: How the Overuse of Antibiotics is Fueling our Modern Plagues. Henry Holt and Co., New York.

Bowler, P.J., 2003. Evolution: The History of an Idea. University of California Press, Berkeley, CA.

Brady, C., 2009. Elizabeth Blackburn and the Story of Telomeres: Deciphering the Ends of DNA. MIT Press, Cambridge, MA.

Brockman, J., 1997. The Third Culture: Beyond the Scientific Revolution. Edge Science Fiction and Fantasy Publishing, Calgary, Canada.

Cassade, A., Gambino, R., 2011. Interactions between Gut Microbiota and Host Metabolism Predisposing to Obesity and Diabetes. Ann. Rev. Med. 62, 361–380.

Chargaff, E., Davidson, J.N., 1955. The Nucleic Acids. Academic Press, New York.

Comfort, N.C., 1999. The real p point is control: the reception of Barbara McClintock's controlling elements. J. Hist. Biol. 32, 133–162.

Cook-Deegan, R., 1994. The Gene Wars: Science, Politics and the Human Genome. W.W. Norton & Company, New York.

Coyne, J.A., 2009. Why Evolution is True. Oxford University Press, Oxford.

Crick, F.H.C., 1966. The genetic code: III. Sci. Am. 215, 55–62.

Crick, F.H.C., 1981. Life Itself: Its Origin and Nature. Simon & Schuster, New York.

Darwin, C., 1859. On the Origin of Species by Means of Natural Selection, or the Preservation of Favoured Races in the Struggle for Life, first ed. John Murray, London.

Dawkins, R., 1976. The Selfish Gene. Oxford University Press, Oxford.

Deamer, D., Szostak, J.W., 2010. The Origins of Life. Cold Spring Harbor Laboratory Press, Cold Spring Harbor, NY.

It's in Your DNA. http://dx.doi.org/10.1016/B978-0-12-812502-1.00021-4

Dobell, C., 1932. Antony van Leeuwenhoek and his "Little Animals". Dover Publications, New York.

Dubos, R., 1976. The Professor, the Institute, and DNA: Oswald T. Avery, His Life and Scientific Achievements. Paul & Company, Baltimore, MD.

Eriksen, M., et al., 2013. The Tobacco Atlas. American Cancer Society, Inc., USA.

Falkow, S., 1975. Infectious Multi-Drug Resistance. Pion Ltd., London.

Finch, J.A., 2008. Nobel Fellow on Every Floor: A History of the Medical Research Council Laboratory of Molecular Biology. Medical Research Council, Cambridge.

Fredricks, D.N., 2013. The Human Microbiota: How Microbial Communities Affect Health and Disease. Wiley-Blackwell, Oxford, UK.

Glynn, J., 2012. My Sister Rosalind Franklin. Oxford University Press, Oxford.

González, M.B.R., Gonzalez-López, J., 2013. Beneficial Plant-Microbial Interactions: Ecology and Applications. CRC Press, Boca Raton, Florida.

Gould, S.J., 2002. The Structure of Evolutionary Theory. Belknap Press of Harvard University Press, Cambridge, MA.

Gregory, T.R., 2005. The Evolution of the Genome. Elsevier, San Diego.

Harari, Y.N., 2012. From Animals Into Gods: A Brief History of Humankind. Kinneret, Zamora-Bitan, Dvir—Publishing House Ltd., Israel.

Henig, R.M., 2000. The Monk in the Garden: The Lost and Found Genius of Gregor Mendel, the Father of Genetics. Houghton Mifflin, Boston.

Herschel, J., 1830. A Preliminary Discourse on the Study of Natural Philosophy. Longman, Rees, Orme, Brown & Green and John Taylor, London, Reissued by Cambridge University Press, 2009.

Huxley, J., 1965. Charles Darwin and his World. Viking Press, New York.

Jablonka, E., Lamb, M.J., 2005. Evolution in Four Dimensions. MIT Press, Cambridge, MA.

Judson, H., 1979. The Eighth Day of Creation: Makers of the Revolution in Biology. Simon & Schuster Inc., New York.

Kanungo, M.S., 1994. Genes and Aging. Cambridge University Press, Cambridge.

Keller, E.F., 1983. A Feeling for the Organism. W. H. Freeman and Company, New York.

Keynes, R., 2001. Charles Darwin's Beagle Diary. Cambridge University Press, Cambridge, UK.

Kornberg, A., 1989. For the Love of Enzymes: The Odyssey of a Biochemist. Harvard University Press, Cambridge, MA.

Lack, D., 1983. Darwin's Finches. Reissued in 1983 by Cambridge University Press.

Lagerkvist, U., 1998. DNA Pioneers and Their Legacy. Yale University Press, New Haven, CN.

Lyell, C., 1830. Principles of Geology: Being an Attempt to Explain the Former Changes of the Earth's Surface, by Reference to Causes now in Operation. John Murray, London.

Lyte, M., Cryan, J.F., 2014. Microbial Endocrinology: The Microbiota-Gut-Brain Axis in Health and Disease. Springer, Heidelberg.

Margulis, L., 1971. Origin of Eukaryotic Cells. Yale University Press, New Haven, CT.

Mayr, E., 1942. Systematics and the Origin of Species. Columbia University Press, New York.

Mayr, E., 2005. One Long Argument: Charles Darwin and the Genesis of Modern Evolutionary Thought. Harvard University Press, Cambridge, MA.

Morgan, S., 2003. Superfoods: Genetic Modification of Foods (Science at the Edge). Heinemann Publishing, Portsmouth, NH.

Mukherjee, S., 2010. The Emperor of All Maladies. A Biography of Cancer. Scribner, New York.

Mukherjee, S., 2015. The Gene. An Intimate History. Scribner, New York.

Mullis, K., 1990. The unusual origin of the polymerase chain reaction. Sci. Am. 262, 56–65.

Mullis, K., 1998. Dancing Naked in the Mind Field. Pantheon Books, New York.

Nussbaum, R.L., et al., 2001. Genetics in Medicine. Saunders, Philadelphia.
Olby, R., 1994. The Path to The Double Helix: Discovery of DNA; first published in October 1974 by MacMillan, Kowloon, Hong Kong, with foreword by Francis Crick; revised in 1994, with a 9-page postscript.
Oparin, A.I., 1938. The Origin of Life on Earth. The Macmillian Company, New York, This a translation of the book first published in Moscow. The book is available in paperback by Dover, NY (1953).
Otles, S., 2013. Probiotics and Prebiotics in Food, Nutrition and Health. CRC Press, Boca Raton, Florida.
Pasachoff, N., 2006. Barbara McClintock, Genius of Genetics. Enslow Publishers, Inc., Berkeley Heights, NJ.
Patent, D.H., 2001. Charles Darwin: The Life of a Revolutionary Thinker. Holiday House, New York.
Portugal, F.H., Cohen, J.S., 1977. A Century of DNA. MIT Press, Cambridge, MA.
Quackenbush, J., 2011. Curiosity Guides: The Human Genome. Charlesbridge Publishing, Watertown, MA.
Rabinow, P., 1996. Making PCR: A Story of Biotechnology. University of Chicago Press, Chicago.
Rhodes, R., et al., 2014. The Human Microbiome: Ethical, Legal and Social Concerns. Oxford University Press, Oxford, UK.
Ridley, M., 2006. Francis Crick: Discoverer of the Genetic Code. Harper Collins Publishers, Glasgow, UK.
Rosenberg, E., Gophna, U. (Eds.), 2011. Beneficial Microorganisms in Multicellular Life Forms. Springer, Heidelberg.
Rosenberg, E., Zilber-Rosenberg, I., 2014. The Hologenome Concept: Human, Animal and Plant Microbiota. Springer, New York.
Rutherford, A., 2013. Creation: The Origin of Life; The Future of Life. Penguin Books, London, England.
Sapp, J., 1994. Evolution by Association: A History of Symbiosis. Oxford University Press, New York.
Sayre, A., 1975. Rosalind Franklin and DNA. Norton & Co., New York.
Shine, I., Wrobel, S., 1976. Thomas Hunt Morgan: Pioneer of Genetics. University of Kentucky Press, Lexington, Kentucky.
Smith, J.M., 2005. Genetic Roulette. Yes Books, St. Louis, MO.
Snow, C.P., 2001. The Two Cultures and the Scientific Revolution: The Two Cultures. Cambridge University Press, London, UK.
Stent, G.S., 1971. Molecular Genetics: An Introductory Narrative. W. H. Freeman, San Francisco.
Steven Austad, S., 1997. Why We Age: What Science is Discovering About the Body's Journey Through Life. John Wiley & Sons, London.
Sussman, S., 1989. Big Friend, Little Friend: A Book about Symbiosis. Houghton Mifflin, Boston, MA.
Swartz, J., 2008. In Pursuit of the Gene. From Darwin to DNA. Harvard University Press, Cambridge, MA.
Tudge, C., 2000. In Mendel's Footnotes: An Introduction to the Science and Technologies of Genes and Genetics from the Nineteenth Century to the Twenty-Second. Vintage, London.
Velasquez-Manof, M., 2012. An Epidemic of Absence: A New Way of Understanding Allergies and Autoimmune Diseases. Scribner Publishers, New York.
Wald, G., 1954. The origin of life. Sci. Am. 191, 44–53.
Watson, J.D., 1965. Molecular Biology of the Gene. W.A. Benjamin, Inc., New York.
Watson, J.D., 1968. The Double Helix. Atheneum, New York. In 1998, the Modern Library placed The Double Helix at number 7 on its list of the 20th century's best works of non-fiction.

Watson, J.D., 2007. Recombinant DNA: Genes and Genomes: A Short Course. W.H. Freeman, San Francisco.

Watson, J.D., Tooze, J., 1981. The DNA Story: A Documentary History of Gene Cloning. W.H. Freeman & Company, San Fransisco.

Weaver, S., Morris, M., 2003. An Annotated Bibliography of Scientific Publications on the Risks Associated with Genetic Modification. Victoria University Press, Wellington, NZ.

Wells, S., 2002. The Journey of Man: A Genetic Odyssey. Princeton University Press, Princeton, NJ.

Wilkins, M., 2004. The Third Man of the Double Helix: The Autobiography of Maurice Wilkins. Oxford University Press, Oxford, UK.

Williams, G.C., 1966. Adaptation and Natural Selection. Princeton University Press, Princeton, NJ.

Woese, C., 1967. The Genetic Code: The Molecular Basis for Genetic Expression. Harper & Row, New York.

Wostmann, B.S., 1996. Germfree and Gnotobiotic Animal Models: Background and Applications. CRC Press, Boca Raton, Florida.

Yarus, M., 2013. Life From an RNA World: The Ancestor Within. Harvard University Press, Cambridge, MA.

Index

A

B

C